Adaptive Analog VLSI
Neural Systems

Adaptive Analog VLSI
Neural Systems

M.A. Jabri

R.J. Coggins

and

B.G. Flower

SEDAL, Sydney University
Electrical Engineering, Australia

CHAPMAN & HALL

London · Glasgow · Weinheim · New York · Tokyo · Melbourne · Madras

Published by Chapman & Hall, 2-6 Boundary Row, London SE1 8HN, UK

Chapman & Hall, 2-6 Boundary Row, London SE1 8HN, UK

Blackie Academic & Professional, Wester Cleddens Road, Bishopbriggs, Glasgow G64 2NZ, UK

Chapman & Hall GmbH, Pappelallee 3, 69469 Weinheim, Germany

Chapman & Hall USA, 115 Fifth Avenue, New York, NY 10003, USA

Chapman & Hall Japan, ITP-Japan, Kyowa Building, 3F, 2-2-1 Hirakawacho, Chiyoda-ku, Tokyo 102, Japan

Chapman & Hall Australia, 102 Dodds Street, South Melbourne, Victoria 3205, Australia

Chapman & Hall India, R. Seshadri, 32 Second Main Road, CIT East, Madras 600 035, India

First edition 1996

© 1996 M.A. Jabri, R.J. Coggins and B.G Flower

ISBN 0 412 61630 0

A Catalogue record for this book is available from the British Library

∞ Printed on permanent acid-free text paper manufactured in accordance with ANSI/NISO Z39.48-1992 (Permanence of Paper).

Contents

CHAPTER 1

Overview

Since the pioneering work of Carver Mead and his group at Caltech, analog neural computing has considerably matured to become a technology that can provide superior solutions to many real-world problems.

Many books and technical papers have been written over the last ten years on analog implementations of artificial neural networks in silicon. This book presents chips and systems that have been developed and applied to real-world problems. A strong emphasis is placed on micro-power trainable artificial neural network systems and the book demonstrates how such systems provide adaptive signal processing and classification capabilities.

One of the most exciting applications of micro-power analog artificial neural network systems is in the area of Functional Electrical Stimulation (FES). This area includes systems such as prosthesis devices, cochlear implants, implantable heart defibrillators and retinal implants. Over the next decade, FES system designers will continue to face two fundamental challenges:

- space and miniaturization (volume occupied by the devices)
- power consumption (battery life-time)

Technologies which constitute advances with respect to these two requirements and can offer improved functionality will be of considerable interest and value. It is our belief that micro-power neural networks can make an important contribution towards the establishment of such technologies. Neural network based pattern recognition devices presented in this book demonstrate examples of how complex computation can be achieved with very small area and at ultra-low power consumption.

Analog microelectronic implementations of artificial neural networks offer a number of attractive features:

- they provide an inherent parallelism since computational elements can be highly compact and interconnected;

- in many real-world applications they do not require expensive and bulky analog to digital converters as they can interface directly to analog sensory inputs;

- when implemented using weak inversion CMOS circuit design techniques, the resulting system can operate at very low power which could be very attractive for pattern recognition applications in battery operated devices;

- analog implementations can provide higher operating speeds than digital implementations within their precision and noise margins;

- in fully parallel implementations, they provide a degree of fault tolerance because information is represented in a distributed fashion.

However, the attractive features of analog computation can only be exploited if the following obstacles can be tolerated or overcome:

- accuracy is limited in practice to less than 8 bits which can degrade the performance of learning algorithms;

- noise immunity is a serious design constraint;

- automated analysis and simulation tools of analog circuits are limited when compared to digital simulation and timing verification tools;

- memory is not accurate and is bulky to implement in analog.

This book presents a cross section of efforts to address the obstacles above in three laboratories, the Systems Engineering and Design Automation Laboratory at Sydney University Electrical Engineering, the Adaptive Systems Research Department at AT&T Bell Laboratories and the Neural Networks Group at Bellcore. The book does not aim to cover every technique or implementation methodology for analog neural networks. However, the circuits and algorithms presented here have been used successfully to perform practical tasks and can be used to implement neural computing architectures other than those we cover here.

An important aspect of this book is the linkage between neural network architectures and learning algorithms. Learning algorithms for analog neural networks (also called VLSI friendly learning algorithms) are a challenging research area and critical to the successful application of analog neural computing to real-world applications.

Book roadmap

In Chapter 2, we provide a brief introduction to neural computing. A framework is first introduced, followed by a description of
the perceptron, the multi-layer perceptron and their learning algorithms. This chapter covers sufficient material from the neural
computing field for the understanding of the architectures of most
systems described in this book.

Chapter 3 provides an introduction to CMOS devices and circuits. The basic MOS device is described and the drain current
equation is developed. The cases of strong and weak channel inversion are considered. Basic analog circuits and signal converters are
presented, including digital to analog and analog to digital converters.

Basic neural computing building blocks are presented in Chapter 4. The process of moving from a functional design to an
implementation architecture is described. The trade-offs in device
choice, area, power and system level design are discussed. The design of basic neurons and synapses is presented and the relative
advantages discussed.

In Chapter 5, a simple analog micro-power multi-layer perceptron with off-chip neurons, called Kakadu, is presented. This chapter brings together the theory and elements of VLSI analog ANN
design presented in the previous chapters, particularly the building
blocks described in Chapter 4, in a simple but functional system.
The effects of using nonlinear synapses and linear neurons are discussed. The chapter concludes with a summary of the successful
use of the Kakadu chip in the classification of cardiac arrhythmias.

Training algorithms and strategies for analog artificial neural
networks is the subject of Chapter 6. Sequential, semi-parallel and
parallel weight perturbation learning algorithms are described. The
practical efficiency of these algorithms is measured by extensive experiments using the Kakadu chip on two problems, parity 4 and
intracardiac electrogram classification. Comparisons of perturbation based algorithms with back-propagation are also made in the
context of analog VLSI implementation.

Chapter 7 describes Snake, an analog micro-power artificial neural network system. Snake includes an analog bucket brigade chain,
a multi-layer perceptron and a winner-take-all subsystem. Its inter-
facing to an implantable cardioverter defibrillator is described and
its training and operation as a classifier of cardiac arrhythms at

very low power shows the advantages of such technology in solving real-world problems that could not be solved by traditional techniques under the same power and area constraints.

In Chapter 8, chips implementing integrated (on-chip) learning are described. The chapter shows that analog micro-power pattern recognition systems that can adapt their behavior in real-time without the assistance of external systems are achievable in the short term. One of the chips called Tarana, implements several of the sequential and semi-parallel weight perturbation based learning techniques described in Chapter 6. The chapter also describes analog memory circuits for short term storage of the weights.

Alternative analog weight storage circuits are described in Chapter 9. The chapter compares three storage approaches from points of view of the reliability and practicality and within the context of on-chip learning. An analog to digital and digital to analog-based memory storage circuit is presented and its operation is described.

The design of a switched capacitor based neural network is described in Chapter 10. The neural network is operated in charge transfer mode. Its advantages with respect to implementations in earlier chapters are discussed. One of the main difficulties faced in the design of analog micro-power neural systems like those described in Chapter 7 and Chapter 8 is the design of a linear neuron with variable gain. The design of such a neuron using switched capacitor techniques is presented and its operation is discussed.

Chapter 11 presents the design, implementation and application of a neural network accelerator board based on AT&T's NET32K chip. The NET32K chip, an efficient analog convolution engine, is reviewed and applications of the accelerator board in image processing and optical character recognition are described.

Finally, an accelerator board based on Bellcore's Boltzmann chip is described in Chapter 12. The principles of designing a Boltzmann chip with integrated learning are presented and its integration into a general purpose system is discussed. The chapter also presents applications of the board in pattern recognition and optimization problems.

Acknowledgements

The preparation of this book was greatly simplified by the energetic support of Jian-Wei Wang from Sydney University Electrical Engineering. It was great to receive the support of somebody with a neural computing background, who could also drive LaTeX and FrameMaker. Early versions of some chapters were prepared using FrameMaker and their migration to LaTeX is due to him. Thanks Jian-Wei!

We would like to thank André Van Schaik and John Lazzaro for their helpful comments and suggestions on the draft manuscript of the book. Finally, we would like to acknowledge the support of all past and present members of the Systems Engineering and Design Automation Laboratory at Sydney University Electrical Engineering. In particular, we wish to thank Dr Stephen Pickard, Dr Philip Leong and Edward Tinker for their contributions to analog micro-power computing in our laboratory.

CHAPTER 2

Introduction to neural computing

2.1 Introduction

This chapter describes basic neural computing architectures and learning algorithms targeted for implementation in later chapters. It is intended for the reader seeking an abstract and simple introduction to neural computing architectures and learning algorithms. We first describe a basic neural computing framework, and then review the perceptron, multi-layer perceptron, and associated learning algorithms.

2.2 Framework

2.2.1 Neurons and synapses

In the simplest abstract form, a neuron is a device that performs two functions (see Figure 2.1):

1. summation and
2. transformation

The input to the summation process are signals produced by the synapses feeding into the neuron. A synapse acts as a multiplicative weighting element which has associated with it a weight w, an input signal x and produces the output signal $w \cdot x$. If a neuron has N synapses connected to it, the outcome of the summation process is the value net, where

$$net = w_0 \cdot x_0 + w_1 \cdot x_1 + ... + w_{N-1} \cdot x_{N-1} = \sum_{i=0}^{N-1} w_i \cdot x_i.$$

The transformation function of the neuron operates on the result of the summation. Typical transformation functions are shown in Figure 2.2 and two of the most popular ones are the *sigmoid* function

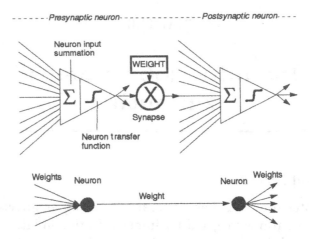

Figure 2.1 *Two schematic representations of neurons and synapses. The upper representation is structural and the lower is abstract.*

$$f(net) = 1/(1 - e^{-net})$$

and the hyperbolic tangent function

$$f(net) = tanh(net).$$

From a microelectronic point of view the $tanh()$ function is attractive because it is easy to implement using transconductance amplifiers. We will see later that some learning algorithms require the transformation function to be differentiable.

2.2.2 Neural networks

A neural network consists of a network of neurons interconnected by synapses. The neurons might be organized in layers where each neuron in a layer is connected via synaptic weights to all the neurons in the next layer. The neurons may also be arbitrarily connected.

To the user, the neural network appears as a 'black-box' with inputs and outputs (Figure 2.3). The inputs and outputs are usually multi-dimensional signals and will be represented by the vectors **x** and **y** respectively. Inside the 'black-box', the input signal **x** is

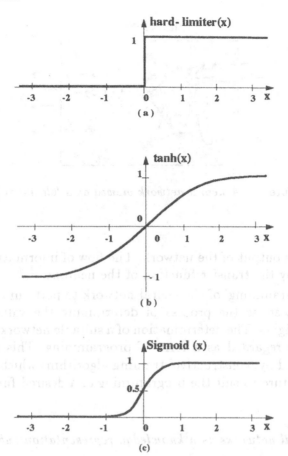

Figure 2.2 *Typical neuron transfer functions.*

processed by the network of neurons to yield the output signal. Hence the output signal **y** is effectively the output of neurons.

Given an input signal, the function of a network is determined by:

- its architecture (topology of the network);
- the values of the synaptic weights;
- the transfer functions of the neurons.

The architecture of the network is related to the target application and the amount of information to be 'stored' in the network. The information is stored in the synaptic weights. The values of the synaptic weights control the transfer of information from the input

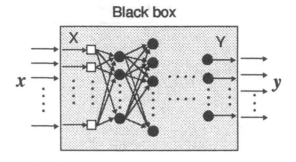

Figure 2.3 *A neural network viewed as a 'black-box'.*

signal to the output of the network. The flow of information is also controlled by the transfer function of the neurons.

The 'programming' of the neural network to perform a function is usually seen as the process of determining the values of the synaptic weights. The determination of a suitable network topology can also be regarded as a form of programming. This would be accomplished by 'constructive' training algorithms which optimize the architecture to suit the programming of a desired function.

2.2.3 Neural networks as a knowledge representation substrate

Once programmed (or trained), a neural network transforms an input pattern of data (signal **x**) into outputs that represent an interpretation (transformation) of the input. The network is therefore a knowledge representation substrate. In comparison to other knowledge representation schemes such as decision trees or rules in a rule-based system, the knowledge is stored in the neural network in a distributed form. That is, an inspection of a single synaptic value does not necessarily provide any recognizable information and it is the (collective) ensemble values of the synapses, in conjunction with the connectivity (architecture) and the transformation applied to them (the neuron functions), which represent knowledge. This parallel and distributed processing [Rumelhart, Hinton and Williams (1986)] is one of the most important attributes which differentiate neural network-based machines from other artificial intelligence knowledge representation schemes.

2.3 Learning

Learning can be separated into two broad paradigms: unsupervised and supervised. Most learning algorithms were developed to cater for a particular architecture. When neural computing is used as a technology (e.g. tool to create a classifier), the separation between architectures and learning algorithms is desirable. However, when neural computing is used as a vehicle to further the understanding of natural intelligence, this separation is considered less acceptable. In this latter context, learning algorithms have to be 'simple', biologically plausible, and cater for varying information processing requirements.

For unsupervised learning, a neural network is only provided with a basic 'organization' rule which is used to adapt its structure in response to input patterns. That is, there is no feedback or correction from the environment (no teacher) in response to the outputs it generates. Hebbian learning [Hebb (1949)] is one of the oldest unsupervised learning rules and was proposed by Donald Hebb in 1949. In its original form, the weights of a network are modified at iteration $n + 1$ according to

$$w_{ij}(n + 1) = w_{ij}(n) + \eta y_i y_j \qquad (2.1)$$

where w_{ij} is the weight from neuron i to neuron j, n is some discrete, arbitrary time increment, and y_i, y_j are the outputs of neurons i and j respectively and η is a learning rate. This rule amplifies weights connecting neurons which respond in the same direction (sign) and attenuates the weights of neurons which respond differently to an input pattern. Since its original formulation, the Hebbian learning rule has been modified to include the additional constraint that weights are limited in magnitude.

For supervised learning, a teacher provides the desired output of a network. In this context, learning can be stated as an optimization problem: find the values of the weights which minimize the error between the actual output and the desired (teacher) values. The perceptron learning procedure and back-propagation (both described below) are examples of supervised learning.

The neural network architectures and learning algorithms described in this book are of the supervised type. The circuit implementations, however, can be used to implement unsupervised learning networks.

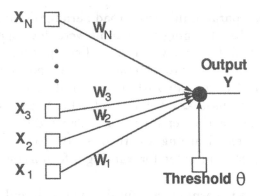

Figure 2.4 *Schematic of a perceptron.*

2.4 Perceptrons

A perceptron is shown in Figure 2.4. Note the inputs x_1, x_2, ..., x_N are pins providing the input signals to the neuron which is shown as a shaded circle. Using the framework described earlier, the perceptron's output y is

$$y = f_h(net) \qquad (2.2)$$

where the function f_h is the hard-limiter

$$f_h(z) = \left\{ \begin{array}{ll} +1 & \text{if } z \geq 0 \\ -1 & \text{if } z < 0 \end{array} \right. \qquad (2.3)$$

and *net* is defined as

$$net = (\sum_{i=1}^{N} x_i \cdot w_i) - \theta \qquad (2.4)$$

where θ is a threshold. The weights are determined through a learning process known as the Perceptron convergence procedure, which was developed by Rosenblatt (1961).

The perceptron was proposed as a pattern recognition and classification device. We will use an example of its operation for a two-input problem. The two inputs x_1 and x_2 are features that describe objects of two classes C_1 and C_2. The classes can be distributed as shown in Figure 2.5. In other words, the output y of

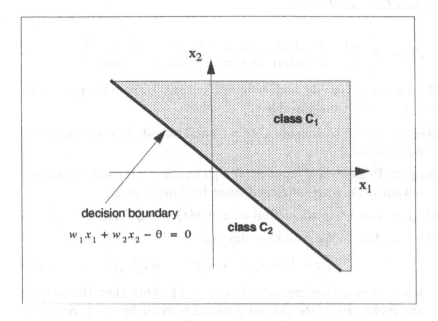

Figure 2.5 *Dichotomy produced by a two-input perceptron.*

the perceptron is positive for any input vector (x_1, x_2) that belongs to class C_1, and is negative for the vectors (x_1, x_2) that belong to class C_2.

On the boundary between the two classes we have

$$w_1 x_1 + w_2 x_2 - \theta = 0 \qquad (2.5)$$

which is the equation of the straight line separating the two classes C_1 and C_2.

In the general N input case, the separation between the two classes is a hyperplane defined by the equation

$$w_0 x_0 + w_1 x_1 + w_2 x_2 + ... + w_{N-1} x_{N-1} - \theta = 0 \qquad (2.6)$$

The important aspect to realise about the separating hyperplane as defined by Equation 2.6 is that it is a linear decision surface.

2.4.1 Perceptron Learning Procedure

In this section we describe the perceptron learning procedure developed by Rosenblatt (1961). We consider here the case where the perceptron is trained to categorise the patterns into the two classes C_1 and C_2 according to:

$$d(t) = \begin{cases} +1 & \text{if pattern at time } t \text{ belongs to class } C_1 \\ -1 & \text{if pattern at time } t \text{ belongs to class } C_2 \end{cases} \quad (2.7)$$

The patterns (inputs and expected outputs) are referenced with respect to the time variable t.

Step 1: Initialise weights $w_i(t = 0)$ and threshold θ to small random values.

Step 2: Present new input $x_0(t), x_1(t), ..., x_{N-1}(t)$ and the desired output $d(t)$, where $d(t)$ is defined by Equation 2.7.

Step 3: Calculate the actual output $y(t)$ using Equation 2.2.

Step 4: Adapt Weights according to

$$w_i(t + 1) = w_i(t) + \eta[d(t) - y(t)]x_i(t) \quad (2.8)$$

where η is a learning rate $(0 < \eta \leq 1)$. Note that the weights are unchanged if the correct decision is made by the perceptron.

Step 5: Repeat the weight corrections for the next pattern by going to Step 2.

The fact that Perceptrons can only create linear decision regions has been argued in the book by Minsky and Papert (1969). Their book and arguments had a negative effect on the research field of artificial neural networks until the work of Hopfield, Rumelhart et al, in the late seventies and early eighties lead to the revival of neural networks as a research field. The work of Rumelhart *et al* [Rumelhart, McClelland and the PDP Research Group (1986)] fostered an understanding of the multi-layer perceptron, which overcomes some of the limitations of the single layer perceptron.

2.5 The Multi-Layer Perceptron

Perceptrons were limited in capability because they could only solve problems that were first order in their nature. If however, additional layers of non-linear neurons are introduced between the input and output layers as in Fig. 2.6, higher order problems such as exclusive-or can be solved by having the 'hidden units' construct or 'learn' internal representations of higher order.

The Multi-Layer Perceptron is the most popular neural network

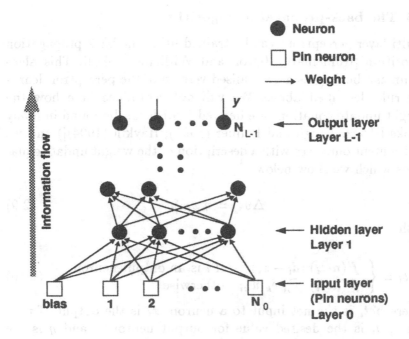

Figure 2.6 *Schematic of a Multi-Layer Perceptron.*

architecture. It is a feed-forward network with the neurons organised in layers. Generally the input layer corresponds to the input to the network as shown in Figure 2.6. The outputs of the network are those of the neurons in the last layer. The neurons usually have associated with them non-linear functions such as the sigmoid and the tanh functions. The neurons of the output layer are often assigned linear functions.

MLPs have been used in a very wide number of applications in pattern recognition/classification and function approximation. In pattern recognition/classification, the input to the MLP corresponds to the features of the object to be classified. Each output of the network is associated with an object class. The recognised object class is that of the strongest output neuron or the strongest of the output neurons that exceed a pre-defined threshold.

In function approximation, the input to the network would be the values of variables of the function and the output of the network would be treated as that of the function to be approximated. The value range of the function to be approximated will be affected by the type (function) of the output neurons.

2.6 The back-propagation algorithm

Multi-layer perceptrons can be trained using the back-propagation algorithm [Rumelhart, Hinton and Williams (1986)]. This algorithm can be seen as a generalised version of the perceptron learning rule described above. We will not reproduce here how the weight update equations are derived (which can be found in many books [Hertz, Krogh and Palmer (1991); Haykin (1994)]) and we will content ourselves with a description of the weight update equations which we show below:

$$\Delta w_{ij} = \eta.x_j.\delta_i \qquad (2.9)$$

with

$$\delta_i = \left\{ \begin{array}{ll} f'(net_i).(d_i - x_i) & \text{if } i \text{ is an output neuron} \\ f'(net_i).\sum_k \delta_k.w_{ki} & \text{otherwise} \end{array} \right. \qquad (2.10)$$

where net_i is the net input to a neuron, x_i is the output of neuron i, d_i is the desired value for output neuron i, and η is the learning rate. This algorithm is a form of optimisation by gradient descent, which is the cornerstone of most neural network training techniques. Note that back-propagation requires the derivative of the neuron output, with respect to the neuron input be computed and the synapse to operate in bi-directional mode.

2.7 Comments

There are many variations on the framework, network architectures and learning algorithms reviewed in this chapter. The self-organising or unsupervised algorithms and networks form a complete area of research in themselves. However, other than the case study in Chapter 12, on a Boltzmann Learning Machine, we will focus on supervised learning exclusively. A wide range of supervised learning algorithms are presented in various forms in the subsequent Chapters with the common theme of gradient descent running through the learning algorithms.

CHAPTER 3

MOS devices and circuits

Graham Rigby

3.1 Introduction

Artificial neural networks could be fabricated in either of the principal integrated circuit technologies - bipolar and MOS. But the rise of interest in the direct implementation of neural networks has come at a time when MOS technology, and CMOS in particular, has become the preferred choice for VLSI. NN's share with large logic systems the requirements for a very low power dissipation per basic function, a high physical packing density and a capability for complex interconnections. These are features of CMOS VLSI. The technology also provides functions such as analog switches and high-quality capacitance which are exploited in many NN designs. On the other hand, in bipolar technology there are circuit functions which arise out of the properties of pn junctions, such as analog multiplication and precision voltage references. These do not come as easily in MOS, though there are ways of providing the functions. There are also technology options which combine MOS and bipolar devices on the same chip. As would be expected, these are more expensive and tend to be used only where there is a very clear performance advantage.

In summary, we focus here on MOS devices for artificial neural networks for the following reasons:

- MOS (CMOS) technology is the predominant production technology and provides the lowest cost per basic function.
- Artificial neural networks typically contain mixed digital and analog functions, both of which are supported by a single technology.
- CMOS development and prototyping services are readily available and processes are well characterised.

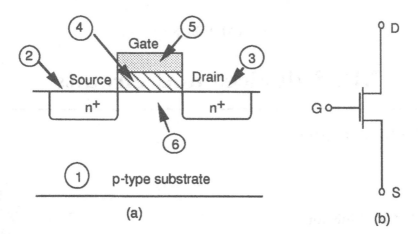

Figure 3.1 *(a) Cross section of an nMOS transistor and (b) circuit symbol of an nMOS transistor.*

Nevertheless, as the rest of this chapter shows, high performance analog functions are often difficult to design in CMOS technology. But we will also see that there are elegant solutions to some of these difficulties.

3.2 Basic properties of MOS devices

In this section, a qualitative description of MOS device operation is given and this is followed by quantitative descriptions and models for use in design. For further details, ther reader may consult [Tsividis (1987); Gray and Meyer (1993); Geiger, Allen and Strader (1990); Mead (1989)]

A simplified cross-section of an n-channel MOS transistor is shown in Figure 3.1(a), with its circuit symbol in Figure 3.1(b). It contains six principal elements. The substrate (1) is single-crystal silicon which is lightly doped p-type. The substrate, or 'bulk' is part of the wafer on which the device is made. The section shown could be packaged as a discrete transistor but is more commonly part of a larger array of devices in an IC. Two regions, (2) and (3) are formed by selective doping with a high concentration of n-type dopant which become the source and drain of the transistor. The space between the source and drain is covered by a very thin layer of insulating silicon dioxide (4) and over this is deposited a conducting layer of heavily-doped poly-crystalline silicon (5). This becomes the gate. Finally, a thin layer of the substrate, immediately below the oxide, is known as the channel region (6), though there is no physical structure within the substrate to distinguish this region.

The source and drain are usually identical in structure, so that there is nothing physically to determine the direction of conduction in the transistor. The distinction is made by the way the device is connected in the circuit.

The acronym MOSFET, standing for Metal-Oxide-Silicon Field-Effect Transistor, is not accurate for a modern transistor, since the gate material, which was metal in earlier devices, has become silicon. The thin insulator may not always be an oxide. Nitrides are sometimes used. But the acronym has become part of the conventional jargon.

The transistor is both a switch and an amplifying device. Large voltage changes on its gate turn it on or off and small gate-voltage variations control current through it. The following explanation applies to the n-channel MOSFET and will be extended later to its complement — the p-channel MOSFET. The type of channel is taken from the doping polarity of the source and drain and is opposite to the doping of the substrate.

Referring again to Figure 3.1(a), the transistor is normally biased with its source to ground or the negative supply and its drain to a more positive voltage. The substrate is also grounded. The OFF state occurs when a zero voltage is applied to the gate. Under these conditions, the path from source to drain is through the p-type substrate material. The pn junction surrounding the source has no voltage across it, while the junction around the drain is reverse-biased by the positive drain voltage. Only a minute junction leakage current can flow. When a voltage of $1V$ or more is applied to the gate, the device turns on as a result of the formation of a thin layer of electrons at the surface of the silicon in the channel region. The gate-oxide-channel combination is a capacitor, with the substrate forming the lower plate. As with any capacitor, the application of a voltage causes charge of opposite sign to build up on the respective plates. In the case of the gate, or upper plate, electrons are displaced, leaving a net positive charge which, of course, is insulated from the rest of the transistor. In the lower plate, which had an excess of holes because of the p-type doping, the holes are first driven out of the channel region and then electrons are attracted in to form a very thin sheet of free electrons at

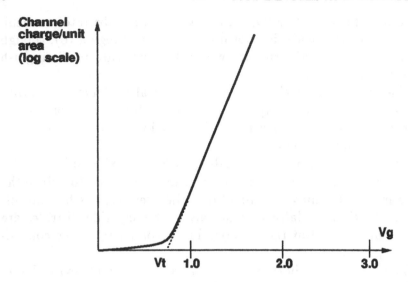

Figure 3.2 *Dependence of the induced channel charge on gate voltage in an nMOS transistor.*

the surface of the silicon. To the source and drain, it appears that the surface of the silicon has become n-type and so the transistor now has a continuous layer of the same type connecting the source and drain. Conduction now occurs along this layer. The amount of current is controlled by the quantity of electron charge induced in the channel, which, in turn is proportional to the gate voltage.

A more detailed examination below shows that the current flow is also influenced by the drain voltage — when this voltage is low. But when the drain voltage is high, the current is almost solely controlled by the gate voltage and this latter mode is the principal one for analog and amplifying applications.

3.3 Conduction in MOSFETs

As stated above, conduction requires the creation of a thin layer of electrons under the gate, which are free to move and which link the source and drain. The full details of this process are beyond the scope of this chapter. It is complex and has been studied intensively for more than 20 years [Sze (1984); Tsividis (1988)].

As the gate voltage is increased from zero, electrons start to accumulate under the gate region. This build up is very gradual until the gate voltage reaches the threshold voltage (V_{to}). Thereafter, the build-up becomes almost proportional to the amount by which the gate voltage exceeds V_{to}. This is illustrated in Figure 3.2 by the solid line. A simplified model of the process, which is widely used to establish the conduction equations, assumes that the source is grounded and deals only with the linear portion of the curve, known as the strong inversion region. Here, however, a more generalised model is presented, which accommodates the case where the source is at an arbitrary potential and also the case where the gate voltage is below the threshold voltage. The latter is known as the weak inversion or sub-threshold conduction region.

This model, which is analogous to the Ebers-Moll model of bipolar transistors, can be applied over the range of bias conditions encountered in analog design. It is known as the EKV model. In what follows, though, a number of basic physical results will be introduced without derivation, in order to keep the treatment concise.[Vittoz (1994)]

Figures 3.3(a) and 3.3(b) establish a set of conventions and parameters which are used in the following analysis. The source, drain and gate voltages are measured relative to the substrate. Current flow I_{ds}, is made up of two components: a forward current I_f flowing from drain to source and a reverse current I_r in the opposite direction. In an n-channel device these currents are carried by electrons. In Figure 3.3(b), the transistor has a channel length L in the direction of conduction and a width W normal to the current flow. The variable x measures the distance of a point in the channel from the edge of the source. In design and layout, W and L are the two parameters controlled by the designer. The others are set by the process technology. Normally, the width is greater than the length, with $W/L = 2$ being typical of a device in a VLSI logic gate while W/L can exceed 100 for some transistors in an op amp.

The charge induced in the channel by the gate voltage is Q_i per unit area and is a function of the gate and channel voltage at any point along the channel. The effective voltage within the conducting channel layer is $V_c(x)$. The gate has a capacitance to the channel of C_{ox} per unit area.

In the strong inversion region, the channel charge is given approximately by:

$$Q_i = -C_{ox}[V_g - V_{to} - nV_c(x)] \qquad (3.1)$$

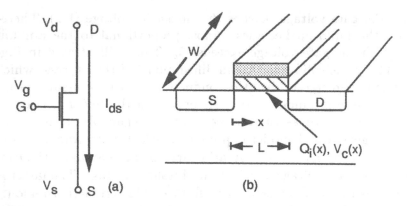

Figure 3.3 *(a) Voltage and current conventions in the MOSFET. (b) Dimensions used in deriving the conduction equations.*

where n is a slope factor determined by the substrate doping and other factors:

$$n = 1 + \frac{\gamma}{2\sqrt{2\phi_f + \frac{(V_g - V_{to})}{n}}} \qquad (3.2)$$

and gamma has a typical value of $0.4V^{1/2}$, while ϕ_f is the Fermi potential in the substrate, with a typical value of $0.6V$. In practice n lies between 1 and 2 and is close to 1 for high values of gate voltage.

In the weak inversion region, there is an exponential relationship between Q_i and the terminal voltages which is approximated by:

$$Q_i = -K_w C_{ox} U_t \exp \left[\frac{V_g - V_{to} - nV_c(x)}{nU_t} \right] \qquad (3.3)$$

in which K_w is a process-dependent coefficient and U_t is the thermal voltage kT/q whose value is $26mV$ at room temperature. Equations (3.1) and (3.3) now describe the curved and linear regions of Figure 3.2, though the transition between them does not have a physically-based analytical expression.

Current flows along the channel in response to a voltage gradient. (The diffusion component is neglected.) In this case, the gradient is given by the derivative of $V_c(x)$ along the channel. Assuming constant current density with x and a uniform current across the width W of the channel, the current flow is expressed as:

$$I_{ds} = -\mu_e Q_i W \frac{dV_c(x)}{dx} \qquad (3.4)$$

in which μ_e is the mobility of electrons in the channel and Q_i is a function of x. This equation is next integrated. For generality, the integration of the right-hand side is broken into two components to separate the dependence of I_{ds} on source voltage and drain voltage, respectively. Thus:

$$\int_0^L I_{ds}\,dx \;=\; I_{ds}L$$

$$\;=\; -\mu_e W \int_{V_s}^{\infty} Q_i dV_c + \mu_e W \int_{V_d}^{\infty} Q_i dV_c \quad (3.5)$$

To draw a parallel with the Ebers-Moll equations, this formulation describes the channel (drain) current in terms of a 'forward' component controlled by the source voltage and a 'reverse' component controlled by the drain voltage. To simplify the appearance of the equations, we also introduce the parameter β defined as:

$$\beta = \mu_e C_{ox} \frac{W}{L} \quad (3.6)$$

Note, also that the product $\mu_e C_{ox}$ is also the conduction factor K used by SPICE and in alternative expressions for the conduction equations.

For the strong inversion case, (3.1) is substituted in (3.5) to give:

$$I_{ds} \;=\; -\frac{\beta}{2n}(V_g - V_{to} - nV_s)^2 + \frac{\beta}{2n}(V_g - V_{to} - nV_d)^2$$

$$\;=\; \beta(V_d - V_s)(V_g - V_{to} - \frac{n}{2}(V_d + V_s)) \quad (3.7)$$

and, for the familiar case of a grounded source, $V_s = 0$, (3.7) reduces to

$$I_{ds} = \beta(V_g - V_{to} - \frac{n}{2}V_d)V_d \quad (3.8)$$

This is the standard non-saturated conduction equation for a MOS-FET in strong inversion, upon which most design theory is based. But there is something incomplete about the equation which needs attention. The last term shows that the curve relating I_{ds} to V_d is an inverted parabola and that when V_{ds} is high enough, I_{ds} would start falling again with increased V_{ds}. But this does not happen in practice. The explanation is that the above analysis carried an implicit assumption that the channel exists along the full distance between source and drain. But if V_d is high, the voltage drop between gate and channel at the drain end will be insufficient to form the channel. The paradox is that the channel does exist at

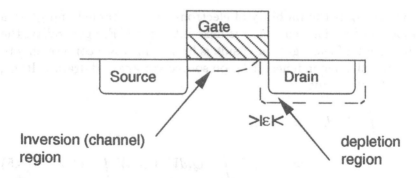

Figure 3.4 *Illustration of the channel gap in a saturated MOSFET.*

the source end, but not at the drain end, yet the device continues to conduct. This is a difficult condition both to understand and analyse, so we will use a qualitative description here, referring to Figure 3.4. A gap forms in the channel so that there is a space of width ε at the drain end. At that point, the voltage within the channel is at the maximum value for the channel to exist, i.e.,

$$V_c(L - \varepsilon) = \frac{V_g - V_{to}}{n} = V_p \tag{3.9}$$

where V_p is known as the pinch-off voltage of the channel. The gap is depleted of carriers and is analogous to the conventional depletion layer at a reverse-biased *pn* junction. In fact, it is part of the depletion layer surrounding the drain-substrate junction. Since depletion layers can sustain a high voltage over a distance typically less than $0.2\mu m$, the gap remains narrow and varies only slightly with drain voltage. Conduction carriers moving along the normal part of the channel reach the edge of this gap and come under the influence of the strong field across the depletion layer, which then sweeps them into the drain region to complete the circuit. This mode of operation is known as saturated conduction and gives the transistor the high output impedance referred to before. Since conduction within the normal part of the channel is still correctly modelled by the theory leading to (3.5), conduction in the saturation region is described by replacing V_d by V_p, as defined by (3.9):

$$I_{ds} = \frac{\beta}{2n}(V_g - V_{to} - nV_s)^2 \tag{3.10}$$

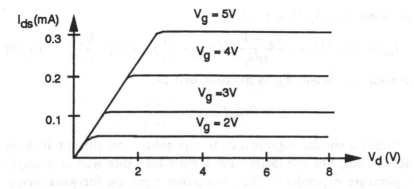

Figure 3.5 *Typical $I_{ds} - V_{ds}$ curves for an n-channel enhancement-mode MOSFET ($V_s = 0$).*

for $V_d > (V_g - V_{to})/n$. This is known as the saturated conduction equation.

For the design of digital circuits, the pair of equations (3.8) and (3.10) provide a sufficiently accurate characterisation of DC behaviour. But in analog design, where the output conductance is critical in determining gain, a further modification is made. As stated above, the depletion layer width in the saturated channel is slightly dependent on V_d. As V_d rises, the depletion layer widens and the effective length of the 'formed' part of the channel is reduced. Since conduction in this part of the channel is inversely proportional to effective length, the result is a rise in I_{ds}, when other parameters are held constant. The exact analysis of this effect is complex, so that it is approximated by a linear term, which is determined empirically. A modified and more accurate form of Equation (3.10) becomes:

$$I_{ds} = \frac{\beta}{2n}(v_g - V_{to} - nV_s)^2(1 + \lambda V_g) \qquad (3.11)$$

The value of λ, the channel-length modulation parameter, varies with channel length and substrate doping. Typical values range from 0.01 per volt for long-channel devices to 0.05 per volt for 1 micron channel lengths.

A combination of (3.8) and (3.11) gives the familiar plot of the $I_{ds} - V_d$ curves for a MOSFET, as shown in Figure 3.5 for the case $V_s = 0$. The above equations apply for strong inversion. To complete the model, the weak inversion case is derived by repeating the integration of (3.5) after substituting the charge density expression of (3.3). The result is:

$$I_{ds} = \beta K_w U_t^2 \exp(\frac{V_g - V_{to}}{nU_t})[\exp(-\frac{V_s}{U_t}) - \exp(-\frac{V_d}{U_t})] \qquad (3.12)$$

A current coefficient I_{do} is next defined as:

$$I_{do} = \beta K_w U_t^2 \exp(-\frac{V_{to}}{nU_t})$$

which has a similar significance to the saturation current in a *pn* junction and, like the latter, has a wide tolerance and is strongly temperature dependent. The conduction equation for weak inversion is now written as:

$$I_{ds} = I_{do} \exp(\frac{V_g}{nU_t})[\exp(-\frac{V_s}{U_t}) - \exp(-\frac{V_d}{U_t})] \qquad (3.13)$$

which is valid for low voltage drops from drain to source.

Finally, it might be noted that an empirical expression has been proposed to describe the transition between conduction in weak and strong inversion. [Vittoz (1994)] This is given in the following equation, though the transition region is not modelled in the following sections of this chapter.

$$\begin{aligned} I_{ds} &= I_{do} \ln^2[1 + \exp(\frac{V_g - V_{to} - V_s}{2U_t})] \\ &\quad - I_{do} \ln^2[1 + \exp(\frac{V_g - V_{to} - V_d}{2U_t})] \end{aligned} \qquad (3.14)$$

3.4 Complementary MOSFETs

So far, the basic theory has been developed around the normally-off nMOS enhancement transistor. There are three other types of interest — a normally-on nMOS device (the depletion transistor) and the two corresponding *p*-channel transistors. Depletion-mode devices are not dealt with here and are not available in most fabrication technologies. But the pMOS enhancement transistor, whose channel current is carried by holes, is equally important as the nMOS device. As Figure 3.6 shows, the structure and circuit symbol are the complements of the nMOS case and the characteristic equations also take the same form, with appropriate sign changes. When nMOS and pMOS devices are combined in circuits, the result is complementary MOS, which has become the mainstream technology for VLSI and a wide range of simpler digital and analog functions. Though CMOS has only occupied this position for less

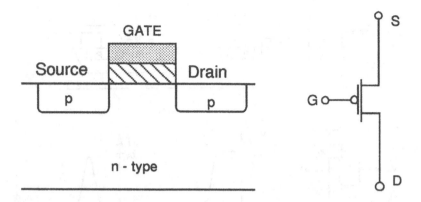

Figure 3.6 *Cross-section and circuit symbol of a pMOS transistor.*

than 10 years, it actually dates back to the early 1970's when RCA started introducing commercial CMOS products. Its slow adoption was partly due to its more complex technology and circuit design, so that its micropower properties were exploited in wristwatch circuits for many years, but not in other high-volume applications. A combination of improved manufacturing technology and the increased importance of minimising power in VLSI led to the widespread use which it now enjoys.

Since n and p-channel devices require substrates of opposite doping, a process step must be added to create a local environment which is correct for the respective devices. If the starting material is n-type, local p-wells are produced by doping — or vice-versa. Both options for starting material are found in production, since there are slight differences in device properties which lead to one being preferable for a given application. Figure 3.7(a) shows a cross section of a complementary pair fabricated in n-well technology.

A CMOS pair of devices connected as in Figure 3.7(a) forms a logic inverter, whose schematic is shown in Figure 3.7(b). This is the simplest building block of digital systems which also has useful analog properties. Figures 3.7(c) and (d) show two DC charactersistics of the inverter which explain the special properties of CMOS. The input-output transfer curve has three regions. At the left, when the input is low, $M1$ is off and $M2$, being a pMOS transistor with a negative V_g, is on. With no external load, the output is connected to V_{dd} by a conducting path, through which no current flows. The output voltage is therefore exactly at V_{dd}.

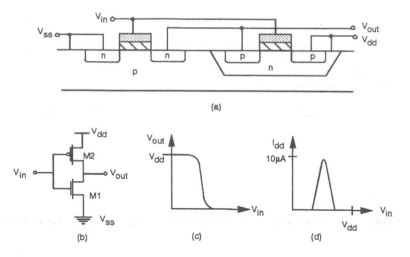

(a)

(b)   (c)   (d)

Figure 3.7 (a) Cross-section and (b) schematic of a CMOS pair connected as an inverter. (c) Static DC transfer characterisitic of a CMOS
inverter. (d) Typical supply current as a function of input voltage.

At the right-hand end of the curve, the opposite happens, with
$M1$ on and an output exactly at ground. This property of having
a full output logic swing is one of the unique properties of CMOS.
The central region corresponds to the transition between the two
levels. The position of this transition is controlled by the relative
conductivity of the two transistors. Since the carrier mobility for
pMOS transistors is lower than for nMOS by a factor of 2 to 3,
the pMOS transistor is frequently designed with a W/L which is
larger by the same ratio. When this is done, the transition region
becomes centred between V_{dd} and ground. Figure 3.7(d) illustrates
the best-know property of CMOS logic - namely, that supply current is only drawn within the transition and not when it is in either
of its stable logic states. In truth, there is a minute current, of the
order of picoamps, due to junction leakage. But a VLSI logic chip,
with more than 100,000 gates, will draw only microamps of supply
current when in its static state.

3.5 Noise in MOSFETs

There are several second-order effects in the physics of MOSFETs,
beyond the channel-length modulation effect already described.
These include noise, the modification of threshold voltage by chan-

nel length and width, and saturation velocity effects in short-channel devices. These are modelled in advanced circuit simulators (eg the Level 3 and Level 4 models in SPICE). Of these effects, noise is probably of greatest interest in analog design and is treated briefly here.

There are two principal noise sources in MOSFETs. [Geiger (1992); Franca and Tsividis (1994)] The first, which applies for normal conduction in the strong inversion mode, is due to the resistive nature of the channel. Analysis of the physics of this process in strong inversion gives a value for the effective noise current in the channel of:

$$I_{dn} = \sqrt{\frac{8}{3}g_m nkT} \qquad V/\sqrt{Hz} \qquad (3.15)$$

where g_m is defined in the next section. In weak inversion, the channel noise consists primarily of shot noise and is approximated by the current generator:

$$I_{dn}|_{(weak)} = \sqrt{2qI_{ds}} \qquad (3.16)$$

Fluctuations in carrier trapping at the channel surface produce a 'flicker' noise which has a spectral density inversely proportional to frequency. This process cannot be analysed exactly and so is approximated by an equivalent input noise voltage to the device:

$$V_{gn} = \sqrt{\frac{4kTr}{WLf}} \qquad (3.17)$$

The parameter r is determined empirically and is very process dependent, with values ranging from $0.02 - 2Vs/Am^2$.

When these two noise processes are superimposed, the result appears as illustrated in Figure 3.8. The noise corner frequency, where the contributions from conduction and flicker noise are equal occurs typically at higher frequencies in MOS circuits than bipolar circuits and is commonly in the 1000–5000 Hz range. Equation (3.15) and (3.17) show that the designer has some control over noise performance through the choice of W/L and bias currents, but it is also the case that some process technologies produce lower noise devices than others.

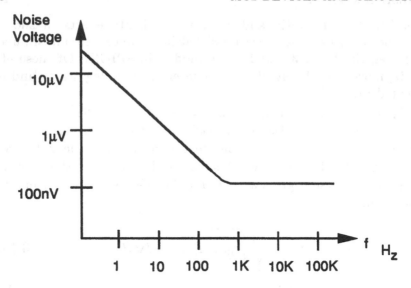

Figure 3.8 *Typical noise output of an MOS transistor.*

3.6 Circuit models of MOSFETs

The results of the previous sections can now be brought together into a set of small-signal circuit models which provide first-order predictions of performance for analog circuits and which also provide insight into the effects of design choices on performance.

Capacitance has been mentioned only indirectly so far. But capacitance exists, of course, between any pair of elements in the MOSFET structure. In particular, the gate has capacitance distributed along it. For convenience, the total is split into two lumped capacitors, one to the source and one to the drain. Also, since source and drain are separated from the substrate by reverse-biased pn junctions, there is capacitance associated with these. The first and more general model is shown in Figure 3.9. The substrate is shown as a separate component, with C_{ss} and C_{ds} modelling the source and drain junction capacitances. C_{gs} and C_{gd} model the two components of the gate capacitance, while C_{ds} models the direct drain-source capacitance, which can usually be neglected.

The gate is assumed to have infinite shunt resistance and is simply a node to control the principal transconductance, g_m. The value of g_m is calculated from (3.11) for a transistor in saturation and can be expressed in several forms to bring out its dependence on

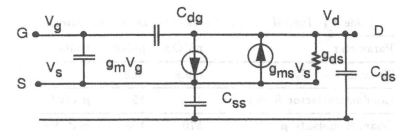

Figure 3.9 *General small-signal MOSFET circuit model.*

other variables:

$$g_m = \frac{\partial I_{ds}}{\partial V_g}$$

$$= \frac{\beta}{n}(V_g - V_{to} - nV_s)(1 + \lambda V_d)$$

$$\simeq \sqrt{\frac{2\beta I_{ds}}{n}}$$

$$\simeq \sqrt{\frac{2K I_{ds}}{n} \cdot \frac{W}{L}} \qquad (3.18)$$

These equations demonstrate the square-root dependence of g_m on bias current and W/L ratio. As a consequence, large changes in geometry or current must be made to produce modest changes in g_m.

In weak inversion, the g_m is derived from (3.13):

$$g_m|_{(weak)} = \frac{I_{ds}}{nU_t} \qquad (3.19)$$

This expression is almost identical to the g_m value for a bipolar transistor and explains one of the interests which this mode holds for analog designers.

The second transconductance element models the source voltage effect on drain current and can be seen from (3.11) to be n times the value of the gate voltage effect.

$$g_{ms} = ng_m \qquad (3.20)$$

for both strong and weak inversion.

Finally, the output conductance element, g_{ds} takes on different forms for non-saturated and saturated conduction. For the former,

Table 3.1 *Typical CMOS Process and Device Parameters.*

Parameter	nMOS	pMOS	Units
Threshold V. V_{to}	+0.8	-0.8	V
Conduction factor $K = \mu Cox$	46	15	$\mu A/V^2$
Channel mobility μ	570	190	cm^2/Vs
Gamma γ	0.7	0.4	$V^{1/2}$
Channel len. mod (λ)	0.01	0.03	V^{-1}
C_{ds}, C_{ss}	0.2	0.5	$fF/\mu m^2$
C_{gd}, C_{ds}	3.5	3.5	$fF/\mu m^2$

it is derived from (3.7):

$$g_{ds}\big|_{(non-sat)} = \frac{\partial I_{ds}}{\partial V_d} = \beta(V_g - V_{to} - V_d) \qquad (3.21)$$

For conduction in saturation, (3.11) is used to give:

$$g_{ds}\big|_{(sat)} = \lambda I_{ds} \qquad (3.22)$$

In theory, (3.22) should also describe the output conductance in weak inversion. In practice, the conductance will be somewhat higher in weak inversion.

Table 3.1 sets out values of the constants introduced above which are typical for a 1.2 micron double-poly CMOS process.[Weste and Eschragian (1993)]

From these, and using the previous equations, a set of parameters, to match the model of Figure 3.9 can be calculated. We will use as our example an nMOS and a pMOS transistor, each with a W/L ratio of 5 and operating at a bias current of $100\mu A$. The channel length is taken as 2 micron and the devices are assumed to be saturated. The parameters are listed in Table 3.2.

The differences in g_m between the two reflect the difference between electron and hole mobility, while the smaller differences in capacitance and output conductance reflect differences in the local doping levels.

The model of Figure 3.9 may now be simplified for more common use. In many circuit connections there is no small-signal voltage

Table 3.2 *Parameter values for the circuit model of Figure 3.9*

Element	nMOS	pMOS	
C_{gs}	28	28	fF
C_{gd}	5	5	fF
C_{ds}	8	8	fF
C_{ss}	8	8	fF
g_m	196	112	$\mu A/V$
g_{ms}	235	134	$\mu A/V$
g_{ds}	1	3	$\mu A/V$

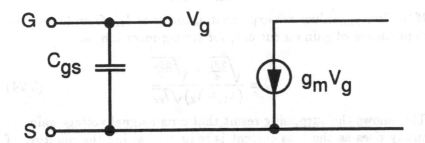

Figure 3.10 *Simplified common-source MOSFET small signal circuit model.*

difference between source and substrate, so that both may be common in the model. Also, two capacitances may be neglected to produce the simplified model of Figure 3.10.

3.7 Simple CMOS amplifiers

A large number of CMOS amplifier configurations have appeared in commercial and research data, demonstrating amongst other things, that there is no ideal amplifier circuit. So the scope here is restricted to the versions likely to be useful in neural network circuits. Moreover, recent trends are towards lower supply voltages for both analog and digital circuits. Designing good analog functions for $5V$ supplies is quite difficult and the movement to $3V$

supplies makes the task harder. So the amplifier circuits most useful in the future will usually be simpler and will tend not to have high 'stacks' of devices between supply and ground.

3.7.1 Inverter-based amplifiers

The conventional CMOS inverter, with its input biased near midrail, acts also as a simple amplifier.[Haskard and May (1988)] Its schematic and DC transfer characteristic were shown in Figure 3.7. It may be modelled around this bias point by using simple small-signal models for the two transistors, as in Figure 3.11. Since both sources are grounded, the g_{ms} sources are omitted. With no external resistive load, the small-signal voltage gain is written as:

$$A_v = \frac{g_{m1} + g_{m2}}{g_{ds1} + g_{ds2}} \qquad (3.23)$$

If the two g_m values are expressed in terms of I_{ds}, from (3.18), the dependence of gain on current, for strong inversion, is:

$$A_v = \frac{\sqrt{\frac{2\beta_1}{n}} + \sqrt{\frac{2\beta_2}{n}}}{(\lambda_1 + \lambda_2)\sqrt{I_{ds}}} \qquad (3.24)$$

This shows the surprising result that small-signal voltage gain actually rises as the bias current is reduced, raising the question of what happens when the current falls to zero. Unfortunately, the result is not an infinite gain, since the devices enter weak inversion mode and the model changes. The gain expression also shows a gain dependence on the W/L ratio, though the dependence is weak.

In weak inversion, the g_m expression from (3.19) is used. As stated previously, the g_{ds} values may be higher than the model predicts, so that the following may be a high value for the gain:

$$A_v|_{(weak)} \simeq \frac{2}{(\lambda_1 + \lambda_2)nU_t} \qquad (3.25)$$

This simple circuit has a useful gain, as the following numerical example shows.

The quiescent current in the transition region depends on the supply voltage and the transistor geometry. Let us take $(W/L) = 10$ for M_1 and $(W/L) = 30$ for M_2. This will match the g_m values of the two transistors. A low supply voltage is assumed, with

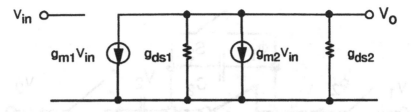

Figure 3.11 *Circuit model for a CMOS inverter operated as a small signal amplifier.*

$V_{dd} = 3V$. From Table 3.1 we calculate:

$$\beta_1 = 460$$

$$\beta_2 = 450$$

The quiescent current $I_{ds}(Q)$ is calculated from (3.10) with $V_s = 0$ and $n = 1.2$.

$$I_{ds}(Q) = \frac{460}{2.4}(1.5 - 0.8)^2 = 94\mu A$$

With the λ values also from Table 3.1, the gain from (3.24) is:

$$A_v = 134$$

The corresponding value for weak inversion is, from (3.25):

$$A_v = 1600$$

Though this high gain makes weak inversion operation appear to be attractive, it should be remembered that this type of circuit does not give a stable bias point for weak inversion and, in fact, the condition would only be achieved with a supply voltage less than $2V_{to}$.

A general limitation of the inverter-based amplifier is that it does not handle differential signals and has a poorly defined input bias voltage. If the amplifier is used in a clocked system, however, an extension is possible using switched-capacitor techniques — which have been widely used in monolithic filters. Referring to Figure 3.12(a), capacitors and switches have been added to the basic inverter.[Haskard and May (1988)] Each switch symbol represents a complementary pair of MOSFETs whose gates are driven by complementary clock voltages. The capacitors are low value ($< 5pF$) components formed on the chip near the inverter. In clock phase ϕ_1, the left side of C_1 is grounded while the input and output

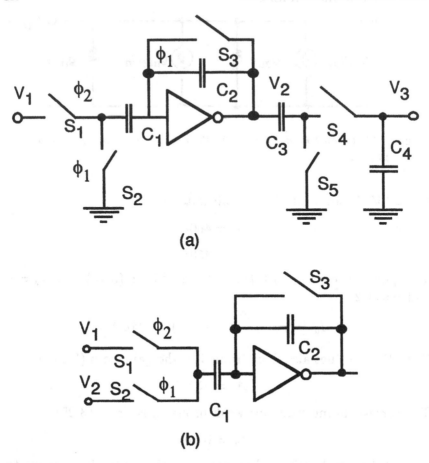

(a)

(b)

Figure 3.12 *Switched-capacitor inverting amplifier for (a) single-ended inputs and (b) differential inputs.*

of the inverter are shorted. C_2 is discharged and the two inverter voltages are forced to the Q-point value of Figure 3.7(c). The output node via C_3 is also grounded. On clock phase ϕ_2, the left side of C_1 rises to V_1, while its right side is held close to the Q voltage by negative feedback through C_2. To conserve charge, therefore, V_2 must fall by $-V_1 C_1/C_2$ and this voltage change is coupled via C_3 to the output node. Switch S_4 then stores the new voltage on C_4.

Like the original inverter, this is still a single-ended amplifier, but the signals are now referenced to ground and the offset voltage is due to second-order effects, namely, finite gain in the inverter and charge injection from the MOS switches. The voltage gain is now defined in terms of a capacitor ratio, which can be held to an accuracy of 2% or better. A further input modification, Figure 3.12(b), converts the circuit to a differential amplifier. The voltage gains for the two circuits are thus:

$$Version(a): \quad V_3 \quad = -\frac{C_1}{C_2}V_1 \qquad (3.26)$$

$$Version(b): \quad V_3 \quad = \frac{C_1}{C_2}(V_1 - V_2) \qquad (3.27)$$

Though the switching adds complexity to the simple inverter circuit, this amplifier is capable of operation from lower supply voltages than the more conventional circuit that follow.

3.7.2 Differential amplifier structures

A fundamentally important differential gain stage, which forms a building block for op amps and comparators, is shown in Figure 3.13. [Gray and Meyer (1993); Geiger Allen and Strader (1990)] The input differential pair M_1, M_2 are matched to each other and are supplied by the common current source I_1. Like the first circuit, this contains no resistors and the load is provided by the pMOS pair M_3 and M_4, which are connected as a current mirror. This has the property that if M_3 and M_4 have equal geometry, a drain current of I_3 flowing in M_3 sets up an equal current I_4 in M_4. The purpose of the mirror is two-fold. It provides a high-impedance load for M_2 at the single-ended output (V_o) and it replicates the signal current from M_1 and directs it also to the output node, thus doubling the effective transfer function of the circuit. This technique is known as an 'active load' and has been adapted from a similar structure in bipolar differential amplifiers.

Unlike the pMOS transistor in the inverter-amplifier, the active load acts as a unity-gain current amplifier, rather than a voltage amplifier and may be modelled by a controlled current source equal to the signal current in M_1 and shunted by the output conductance of M_4. So, after modelling the circuit on a similar basis to Figure 3.11, the differential voltage transfer function can be shown to be:

$$\frac{V_o}{V_1 - V_2} = \frac{g_{m1}}{I_1(\lambda_1 + \lambda_4)} \qquad (3.28)$$

Since this has a similar form to (3.23), we conclude that this amplifier also has a voltage gain in the $100 - 300$ range, though it

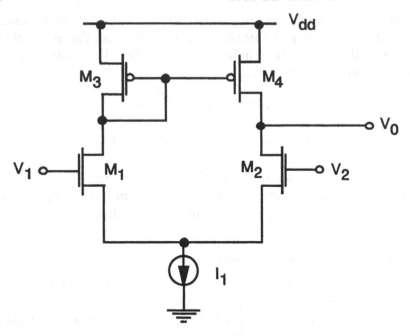

Figure 3.13 *Differential pair with active load.*

requires no switching to produce differential operation. Since it has a high output impedance, the more 'natural' transfer function for the circuit is a transconductance, so that this circuit and its many variants are also known as operational transconductance amplifiers (OTA).

The circuit of Figure 3.13 may also be fully inverted so that a pMOS pair becomes the input. Three small changes result from the inversion. The most obvious is in the common-mode range, i.e. the range of input levels over which the circuit will amplify the difference between the two input voltages. The nMOS input amplifier handles signals up to levels close to the positive supply, but at the low end the range is limited to a voltage greater than the V_{gs} of M_1 and M_2, i.e., about $1V$ above ground. For the pMOS input, the range is displaced downwards. The second effect is on noise. Both types of MOSFETs have a low-frequency $1/f$ noise component which is related to interactions between carriers and trapping centres lying close to surface of the channel. In pMOS transistors, the channel tends to lie deeper in the silicon where the trap density is lower. On the other hand, the g_m of pMOS

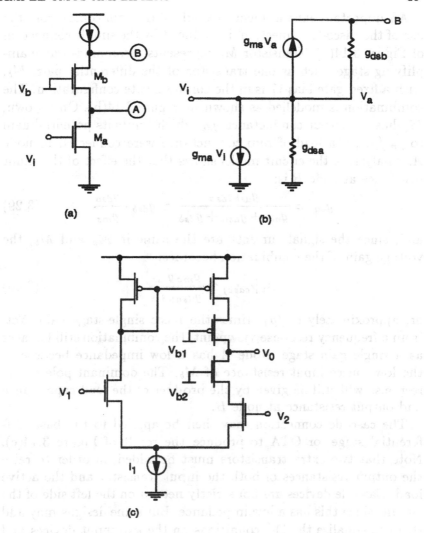

Figure 3.14 *Cascode connections: (a) the basic concept, (b) circuit model of the cascode connection, (c) an OTA with cascode output.*

transistors is a factor of 2 – 3 lower than nMOS transistors under the same bias conditions, so that a larger device is required to deliver the same g_m.

The evolution from the basic circuit to higher performance amplifiers follows two possible paths from here — one towards a higher gain single-stage amplifier and the other towards multi-stage versions. We will consider them in that order.

An important enhancement of single-stage gain comes from the use of the cascode connection, introduced by the simple connection of Figure 3.14(a). Transistor M_a represents a common-source amplifying stage, such as one transistor of the differential pair. M_b, with a fixed gate bias V_b is in the common gate configuration. The combination is modelled as shown in Figure 3.14(b). On its own, M_a has an output conductance g_{ds} which limits its potential gain to g_m/g_{ds}, if a load of zero conductance were connected to node A. Analysis of the circuit model shows that the effect of the shunt resistance at node B is:

$$g_{out} = \frac{g_{dsb}g_{dsa}}{g_{ms} + g_{dsa} + g_{dsb}} \simeq g_{dsb} \cdot \frac{g_{dsa}}{g_{ms}} \qquad (3.29)$$

and, since the signal currents are the same in M_a and M_b, the voltage gain of the combination becomes:

$$A_v|_{(casc)} \simeq \frac{g_{ma}g_{ms}}{g_{dsa}g_{dsb}} \qquad (3.30)$$

or, approximately g_m/g_{ds} times the basic single-stage gain. Yet, from a frequency response viewpoint, the combination still behaves as a single gain stage. Node A has a low impedance because of the low source input resistance of M_b. The dominant pole of the response will still be given by the product of the load capacitance and output resistance at node B.

The cascode connection may then be applied to the basic differential stage, or OTA to produce the result of Figure 3.14(c). Note that two extra transistors must be added, in order to raise the output resistances of both the input transistor and the active load. Cascode devices are not strictly needed on the left side of the circuit, since this has a low impedance. But some designs may add them to equalise the DC conditions on the two input devices and thereby minimise the input offset.

The drawback of this approach comes with low supply voltage operation. If the current source is realised with a single transistor, there is a stack of 5 devices between V_{dd} and ground. Even if only $0.5V$ is dropped across each device, the best case signal swing at the output reduces to $2.5V$ less than the supply voltage.

An important variation of this circuit, the folded cascode helps to overcome this limitation and is illustrated in Figure 3.15. The signal currents from M_1 and M_2 are now diverted downwards through M_x and M_y, by virtue of the two new current sources. A moment's examination shows that there is still a stack of five transistors be-

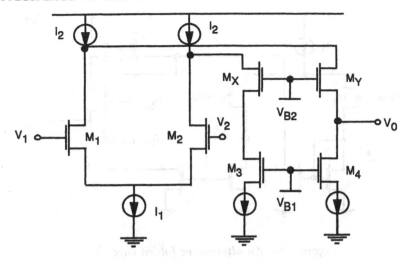

Figure 3.15 *The folded cascode.*

tween V_{dd} and ground, but the circuit decouples the output swing from the input transistors so that both signals can swing to within $1V$ or less from the rails.

A similar result can be obtained simply by using current mirrors to reflect the signal current into the cascode section and an example is illustrated in Figure 3.16. This example uses a pMOS input pair, but could be designed for nMOS as well. Since the current sources of Figure 3.15 are not needed, it would appear that this circuit may be preferable. In fact the differences between the two are second-order. In the second circuit, the voltage at nodes A and B are at the V_{gs} of M_x and M_y, thus restricting the input common mode range. But the corresponding voltage limit in Figure 3.15 is the V_{sat} of the current source transistors, which is less than V_{gsd} by the value of V_t — that is 16% of V_{dd} in a 5 volt circuit.

3.8 Multistage op amps

The design of the 741 family of bipolar op amps, more than 20 years ago, established a classic architecture for a high performance amplifier. This was adopted into CMOS op amp design, where it also produces attractive features. The essential elements of the architecture of multistage op amps are shown in Figure 3.17(a), with a practical and simple realization in Figure 3.17(b). [Gray and Meyer (1993); Geiger, Allen and Strader (1990)] The features of

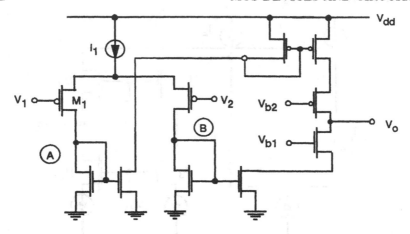

Figure 3.16 *An alternative folded cascode.*

the two-stage plus buffer structure are: a differential input stage,
a single-ended high gain second stage and a unity-gain load giving output stage. The input stage is designed according to the
principles in the previous sections. The second stage, as well as
providing gain, has to solve a common problem in the applications
of op amps, namely, that they be stable in a unity-gain feedback
configuration. This requires a gain transfer function with one very
dominant pole, so that the phase shift at the unity-gain frequency
is much less than 180 degrees. This is achieved by the compensation
capacitor C_c, connected as a Miller-effect capacitor in the second
stage. The third, or buffer, stage provides a low output impedance
and its design problems relate to the achievement of a large output swing, modest power and sufficient load-driving capability. In
Section 3.9, below, comparisons are made between the suitability
of op amps of this type with the single-stage circuits considered
previously.

The role and choice of the compensating capacitor and its relationship to the gain of the op amp can be analysed as follows.
Figure 3.18(a) shows a typical gain and phase plot for a dominant-
pole gain function in which there are also several non-dominant
poles, as will be the case with any practical amplifier. If the output of the op amp is connected to the inverting input, this also
become the description of the loop-gain function. If A_{v_o} is the
magnitude of the low-frequency op amp gain and the magnitude of
the dominant pole is p_1, the asymptote of the gain curve above p_1

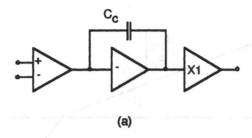

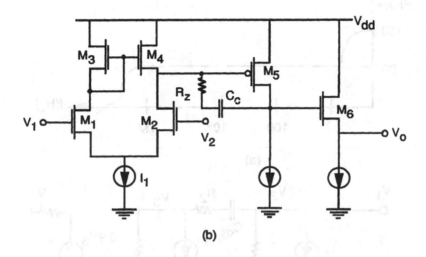

Figure 3.17 *(a) Generic architecture of the two-stage op amp and (b) a simple two-stage op amp design example.*

is approximately $-|A_{v_o}/p_1|$. When the frequency approaches that of the next pole, p_2 the phase angle of the forward gain moves from $-90°$ towards $-180°$ and stability requires that the forward gain must fall below unity before the $180°$ phase shift point is reached. Thus the stability condition is:

$$|p_1| < \left| \frac{p_2}{A_{v_o}} \right|$$
(3.31)

Using the simple circuit model of Figure 3.18(b), an expression may be derived for the approximate transfer function of the op amp (neglecting for now the extra resistor R_z):

$$A_v \simeq \frac{0.9 g_{m1} g_{m5}}{g_{ds2} g_{ds5}} \cdot \frac{1}{1 + S C_c g_{ms} \frac{g_{ds5}}{g_{ds2}}}$$
(3.32)

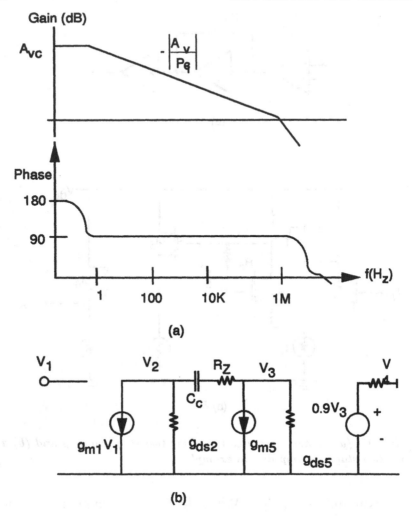

Figure 3.18 *(a) Gain and phase plot for a two-stage op amp. (b) Circuit model for the two-stage op amp.*

When this is combined with (3.31), one sees that the compensation capacitor must satisfy:

$$C_c > \frac{0.9 g_{m1}}{|p_2|} \qquad (3.33)$$

The value of p_2 is not known as a standard result and depends on the detailed circuit design. But experience shows that a value of $2 - 5MHz(12 - 30Mrad/s)$ is common. So if we take the following numerical values:

$$g_{m1} = 500\mu A/V$$

$$|p_2| = 15 Mrad/sec$$

it is seen that a value of $5pF$ would be appropriate for C_c.

A more exact form of (3.32) also shows that the circuit produces a positive real zero in the transfer function. Though not a direct cause of instability, this zero adds to the phase shift in the transfer function and makes compensation more difficult. Thus, the series resistor R_z is included to increase the magnitude of the positive zero as far as possible. In practice it is realised by a long-channel MOSFET.

3.9 Choice of amplifiers

The characteristic of many mixed analog-digital systems is the use of a large number of relatively simple amplifiers or comparators. This is true, for example, in the data converters to be considered in the next section where, in the extreme case, one comparator is used for each analog level in a flash ADC. It is also true in switched capacitor filters, where op amp integrators are used at the rate of one per transfer function pole. In artificial neural networks it is harder to generalise. If the signals being processed are analog then an op amp may form the core of each neuron. Other structures, in which the processing is digital, the attention would be on a few high quality amplifiers in the interfaces. A further factor, which has already been discussed, is the impact of supply voltages below $5V$, which virtually force the adoption of simple amplifier circuits.

Studies of the relationship between op amp gain and circuit performance suggest that a gain of 1000 (60dB) is sufficient for most integrator applications. A few will benefit from gains which are an order of magnitude higher. The previous sections demonstrate that 60 – 80dB gain can be achieved by the one-stage OTA type of circuits. These have the limitation that they are not suitable for resistive load driving but have the advantage that dominant pole compensation is readily provided by the capacitive load and that a separate compensation capacitor is not needed. Thus, in the range of applications dealt with in this book, the OTA is likely to be the first choice. If the gain is adequate, but the supply voltage is low, the choice may move to the switched inverter circuits described near the beginning of Section 3.7. Finally, the circumstances that would favour a full two-stage design would be dictated by the need for resistive load-driving and/or high slew rate into a capacitive load.

3.10 Data converters

At every interface between a digital and an analog system there is some form of analog-to-digital (ADC) or digital-to-analog (DAC) converter. The extent and placement of these interfaces will vary greatly with network architecture since, as later parts of this book show, the processing elements may operate either on digital or analog signals. One common form of NN implementation carries out the multiplication, or weighting of a signal with an analog circuit, but stores the coefficients in digital form. Thus, for our purposes, a selection will be made from the wide variety of possible data conversion techniques. The focus will be on those whose performance matches NN requirements and whose implementation is compatible with the scale and fabrication technology of such networks.

There are some useful generalisations which can be made about data converters to guide our choice and expectations. Since an 8-bit word is a common unit in digital processing, this word length is also a bench-mark for converters. It implies an inherent accuracy or resolution of ±0.4%. This is slightly lower than the matching tolerance achievable in a well controlled IC process, so that 8-bit converters can be readily mass-produced and are widely found both as standard circuit components and are also imbedded in microcontrollers and DSP chips. Increasing the resolution by two more bits to ten takes conventional designs to the limit of high-yield production. Beyond that, other methods are invoked. One is to trim critical components during production — which adds to cost — while the others involve some form of calibration or the use of more clock cycles to carry out a conversion and effectively trade speed for resolution.

Regardless of which method is used, the achievement of high speed and high resolution together is difficult and remains an active frontier in research. Bipolar technology is generally preferred for converters at more than $10M\,samples/sec$, though CMOS performance continues to improve. Data converters, with modest resolution, have been produced in GaAs technology for operation in the $Gsample/sec$ range.

There are two benchmark applications for data converters which are economically important and have stimulated much of the research over the past ten years. The first, digital audio, is a largely solved problem. It requires $16 - 18$ bit resolution at the fairly modest sampling rate of $64k\,samples/sec$. Resolutions of 20–22 bits have been achieved, though at lower bandwidths. The second is driven by the needs of high definition and compressed bandwidth TV. This requires a $10M\,sample/sec$ rate at least and $8 - 10$ bits resolution. This goal has been achieved, though work continues to reduce costs and complexity.

3.10.1 DAC converters

We start with DAC's, because they are usually simpler than ADC's and we start with a one-step converter. [Hoff and Townsend (1979)] The circuit, illustrated in Figure 3.19, consists of a long string of tapped resistors with a tree of analog switches designed to connect one tap only to the output node for a given input code. For an n-bit word there are 2^n levels and, if the positive supply is taken as the reference voltage, $(2^n - 1)$ resistors are required. The required number of switches is about twice this number. The example in Figure 3.19 is for three bits and we can see that if each switch is realised by a complementary MOS pair, there is a further doubling in device count. However, a minor saving in complexity is possible by noting that, unlike most switching circuits, the voltages at most nodes are predictable. Thus, in the top left sector of the tree, voltages will be above $V_{ref}/2$, while they will be below in the lower left. In these sectors, single transistors could be used - pMOS above the centre and nMOS below. To the right, though, the switches must be full CMOS.

The reader is invited to confirm that, for any input binary code, only one conducting path is formed from the resistors to the output buffer.

When a code is presented, a correct output is available within a time equal to the sum of the switch propagation delays and the settling time of the buffer. This delay obviously increases with word length, but typically is in the $100 - 200ns$ range and suits sample rates in the MHz range. The penalty paid is in complexity. A resistor on a chip usually occupies more area than a transistor and area increases with resistance, so that the technique is not suited to micropower design. A final advantage worth noting is that this DAC is inherently monotonic. Though there may be absolute er-

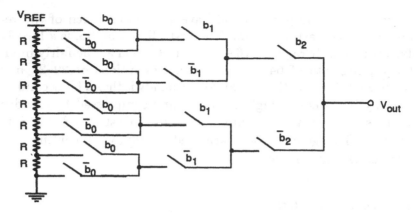

Figure 3.19 *One-step resistor string DAC.*

rors in individual levels, the structure guarantees that the output
for a code of 100, say, is greater than the voltage for 011. This is
not true for some other methods.

The second circuit is also a one-step DAC.[Tsividis and Antognetti (1985)] Like the first example, this uses resistors, this time
arranged in an $R-2R$ ladder, as shown in Figure 3.20(a). But here
the complexity varies linearly with n, rather than exponentially.
The ladder is terminated in a final resistor of R at node n. By
inspection, the resistance to the right of node $(n-1)$ is $2R$. When
the shunt resistor at this node is included, the nodal resistance to
ground is R, so that a voltage division by 2 occurs between node
$n-2$ and $n-1$. By repeating the argument up the ladder, it is seen
that the voltage on each node steps down by a factor of two for
each node from the left hand end where the reference is applied.

There are two possible ways of combining these binary-weighted
voltages into a single output. The voltage mode, which is the more
obvious, must be done without loading the ladder. A CMOS unity-
gain buffer is connected to each node and outputs summed via
switches controlled by the input code, see Figure 3.20(b). Since
the signal path length is similar to that in Figure 3.19, the speed
of this circuit should be similar. However, there are some practical
problems to do with the complexity of the buffer circuit necessary for the overall accuracy and the difficulty of realising accurate
voltage summers.

Operating in the current mode relieves these problems somewhat
— as shown in Figure 3.20(c). The $2R$ resistors are inverted and

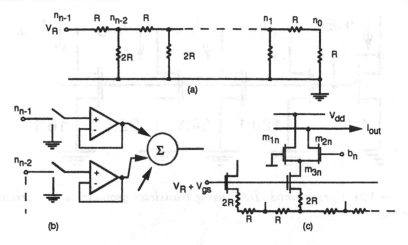

Figure 3.20 *(a) An n-node R-2R ladder; (b) a voltage mode DAC; (c) current-mode DAC.*

used to define currents in the current source of differential current switches. The example combination, $M_{1n} - M_{3n}$ is repeated at each node and if all of the current source gates are biased to $(V_r + V_{gs})$ then the current at the right-hand, MSB end, will be $V_r/4R$ and the currents moving to the right from this point will decrease in binary steps. Here, V_{gs} refers to the operating V_{gs} of each source transistor. But since the I_{ds} in each device is different, so would be its V_{gs}. The solution lies in scaling the W/L ratio of each device down by a factor of 2 as we move to the right. Each source transistor will thus operate at the same current density and hence the same V_{gs}. Above the sources, $M_{1n} - M_{2n}$, for example, act as a current steering switch. The gate of M_{2n} is biased at a fixed voltage V_B, while the right is driven by the logic level corresponding to the complement of the nth bit of the code. Thus, the current for this stage is either drawn from the common output, I_{out} or from V_{dd}. The total current, I_{out} becomes the product of a reference current and the input code and this current may then readily be converted to a voltage in a trans-resistance amplifier.

The scaling of the source transistors described above does carry an area penalty, though. For an 8-bit converter, the source transistor for the MSB should have a W/L ratio 128 times that of the LSB. If the area used is acceptable, this idea can also take a simpler form in which the resistor ladder is removed and the binary weighted currents are simply produced by the scaled areas of the

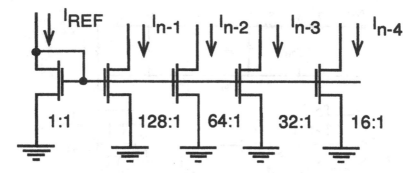

Figure 3.21 *Current-mode DAC using transistor geometries for current scaling.*

source transistors, as in Figure 3.21. This simpler version, relying only on W/L ratios, could be expected to be less accurate than its forerunner, because device matching is difficult when the ratios are high. But in the development of ideas being followed here, it is the first not to use resistors, yet is still a fully static design.

The remaining DAC examples use capacitors. Switched capacitor techniques, which are uniquely associated with MOS technology, were originally developed during the 1980's for analog filter applications. Experience since then has shown them to be surprisingly accurate and they have been adopted into many data converter designs. By nature, they can only be used in clocked systems, so that the following examples differ from the first group, which were all static.

The charge redistribution DAC [Suarez, Gray and Hodges (1975)] (which has an ADC counterpart to be introduced later) requires one cycle of a two-phase clock per sample and has a device count proportional to n, but a circuit area with a 2^n dependence. Figure 3.22 shows one form. For an n-bit code there are n capacitors, with top plates common, arranged in a decreasing binary succession of values. In the first clock phase, all lower plates are grounded and the integrator is reset by switch S_R. In the second clock phase, control logic generates the AND of the clock and each bit of the input code, applying the result to switches $S_1 - S_n$. If a bit is high, the corresponding switch connects the bottom plate to V_r. This injects a charge packet of $V_r C_k / 2^k$ into the integrator which produces an inverted voltage change proportional to the charge. The

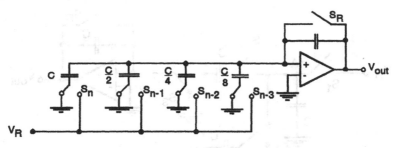

Figure 3.22 *Charge distribution DAC using binary weighted capacitors.*

Table 3.3 *Cyclic DAC Operation*

b_n set	S_2, S_3 up, transferring $V_rC/2$ into integrator. V_a discharged to $V_r/2$ S_2, S_3 switch down	b_n clear	S_4 up, discharging V_a to $V_r/2$
b_{n-1} set	S_2, S_3 up, transferring $V_rC/4$ into integrator. V_a discharged to $V_r/4$	b_{n-1} clear	S_4 up, discharging V_a to $V_r/4$

result is the following transfer function:

$$V_{out} = -V_r(b_n + \frac{b_{n-1}}{2} + \frac{b_{n-2}}{4} + ...) \qquad (3.34)$$

which can be scaled by altering the ratio of the integrator capacitor to the input capacitors.

The second switched-capacitor example, Figure 3.23 trades off speed against complexity. [Franca and Tsividis (1994)] This serial converter requires n two-phase clock cycles per sample, but has a constant complexity with code length. The initial condition has the integrator reset, via S_R. S_1 is closed, charging the first capacitor to V_r, while S_2, S_3 and S_4 ground the the coupling capacitors. For the conversion of the first (MSB) bit and thereafter, S_1 and S_R are open. What follows is described by the Table 3.3.

As the sequence proceeds, it is clear that the integrator accumulates charge proportional to $V_r x$(code value) and hence its output

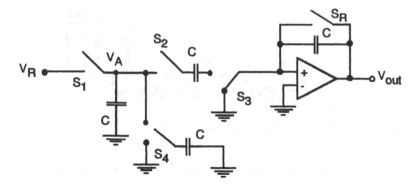

Figure 3.23 *Cyclic DAC based on charge redistribution.*

becomes a negative voltage proportional to this charge. What is more fascinating is that the sequence can repeat an arbitrary number of times and therefore accept an arbitrary long input word. In practice, though, the code length is limited by the accumulation of errors. In the single cycle schemes an error associated with one bit produces a fixed error if that bit is set. So a 1% error in the MSB value, say, simply limits the overall accuracy of the system to 1%. But, in this cyclic scheme, a 1% error in the V_a value after the first cycle becomes a 2% error after the next and so on. So the simplicity of this scheme is offset by the need for very accurate design and technology in realising it. For example, apart from capacitor mismatch, this circuit is also subject to errors due to the finite gain of the integrator op amp. The theory requires the op amp input to be a virtual ground, but if the gain of the real amplifier is 1000, this virtual ground will change in voltage by 0.1% of the integrator output voltage.

3.10.2 Analog-to-digital converters

ADCs generally receive more attention in research papers and books, for several reasons. Conceptually, they incorporate many of the DAC techniques and are therefore more general. Secondly, in mixed systems, there are greater needs for the conversion of analog data, since analog outputs are not needed in many systems. (Radio and television are obvious exceptions.) In this section, some approaches will be omitted as being less relevant to neural networks

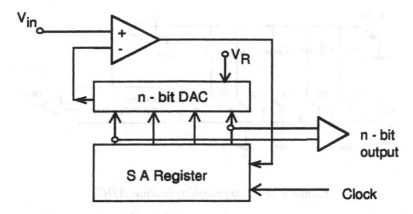

Figure 3.24 *Block diagram of the successive-approximation ADC.*

and we will focus on circuits which are close relatives of the DAC circuits in the previous section.

Successive-approximation converters form a large class, occupying the middle-ground in speed and resolution. A block diagram is shown in Figure 3.24. It is a feedback system in which the return path can be any DAC which produces a result in one clock cycle. Suitable types, from the previous section, include the resistor string, the $R - 2R$ ladder and the current-switching converter. There is also a successive approximation type based on switched capacitors — though the DAC of Figure 3.22 is not directly suitable.

The SA register shown in the diagram consists of a shift-register and separate latch — both of them with the number of bits equal to the resolution of the converter. At the beginning of one cycle, the SAR is cleared and a 1 is loaded into the MSB position of the shift register. On the first main clock cycle, the MSB position of the parallel latch is set, causing the DAC to generate a 1 MSB output. If the comparator determines that this voltage is less than V_{in}, the MSB remains set. But if the DAC output exceeds V_{in}, the latched MSB is cleared. These events occur within one system clock cycle, necessitating the formation of a two or four-phase clock. The next cycle moves the 1 in the shift register along so that it points to the next latch position and the 'testing' of this bit position proceeds.

The process of generating an n-bit output word therefore takes n clock cycles plus the reset phase. The accuracy of the process clearly depends on the accuracy of the DAC in the feedback loop

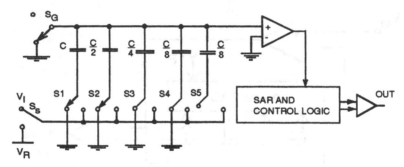

Figure 3.25 *Charge-redistribution ADC.*

and the presence of non-monotonic behaviour in the DAC could give rise to errors greater than one LSB. Overall, the SA architecture — though not simple — is readily designable and represents a good compromise between the many factors influencing ADC design.

The charge-redistribution converter is also a successive approximation type and is described on its own here because it has turned out to be very successful in the middle performance range and is incorporated in a number of one-chip microcomputers which provide analog input data ports. It is illustrated in Figure 3.25 and operates as follows:

At the beginning of the cycle, switch S_G is down, grounding the top plates of all capacitors. S_S is up, and the five numbered switches are to the right, connecting all bottom plates to V_1. Under these conditions, the total charged stored on the capacitor set is seen to be $-2CV_1$.

In step 2 of the cycle, S_G switches up, isolating the top plates, the numbered switches move to the left, grounding the bottom plates and S_S moves to the reference voltage position. At this point, no charge transfer has taken place, the total charge stored is as before, but the comparator input voltage has fallen to $-V_1$.

The remaining steps start a successive approximation process under control of the SA register. S_1 (the MSB) moves to the right, changing the comparator input to $-V_1 + V_R/2$. If this voltage is negative, the bit and the switch remain set. If it is positive, the switch returns. When the next bit down is tested, by operating switch S_2, the comparator input is reduced by $V_R/4$ and this bit is

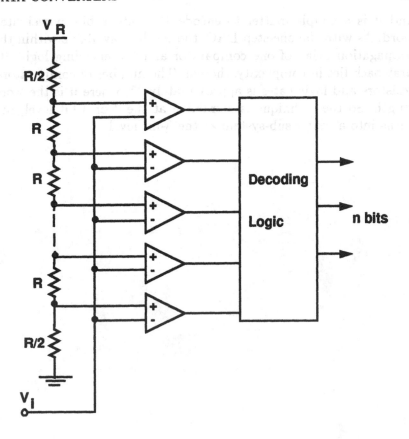

Figure 3.26 *Schematic of a one-step 'flash' ADC.*

then left set or cleared, depending on the sign of the comparator input.

The process then continues until the LSB is tested, each test applying a voltage change which is one half of the previous one.

This chapter concludes with a brief mention of the fastest ADC, the flash converter. It is illustrated in Figure 3.26 and is readily understood. The resistor string sets up a series of evenly-spaced voltages between the reference and ground. As a result of the top and bottom resistors being half the value of the others, each tap voltage lies mid-way between two levels on the scale between V_R and ground. For an arbitrary input voltage, inspection shows that all comparators above a certain level will have low outputs and all below will have high output. This creates a 'thermometer scale' and it is a simple matter to encode this into a binary-weighted word. As with the one-step DAC, the result is available within the propagation delay of one comparator and the encoding logic. Its drawback lies in complexity, though. The number of comparators, resistors and logic gates is approximately 2^n, where n is the word length. So the technique is attractive at the 3 or 4-bit level, but turns into a major sub-system at the 8-bit level.

CHAPTER 4

Analog VLSI building blocks

In this chapter the issues affecting the implementation of the lowest level circuit elements of an analog VLSI neural network are discussed. That is, once a functional specification has been decided upon there are still many design choices available. For most applications these are represented by the trade-off between the resources of circuit area and power consumption against the performance factors of speed, precision and reliability. Directly associated with the trade-off of circuit resources is the trade-off which exists between different VLSI circuit devices. In addition to these low level circuit trade-offs a number of system level implementation issues affect the basic building block implementations. These include system testability, system I/O requirements and system wide manufacturing yield.

4.1 Functional designs to architectures

In this section we describe the choices available in the trade-off of circuit performance versus circuit resources and system level requirements.

4.1.1 Device choice trade-offs

CMOS is currently the most widely used fabrication process for VLSI systems and for the most part is used for the implementation of digital circuits. All CMOS fabrication processes provide a means to fabricate N and P type MOS transistors, diodes, capacitors and resistors. Additionally, processes which provide a second polysilicon layer provide the the opportunity for floating capacitors and floating gate transistors. Further, some processes also provide a special high resistance poly silicon layer for realising resistances as high as a few mega ohms in reasonable area. Here, the discussion of devices will be restricted to widely available processes with

a second poly silicon layer and no special tolerance specifications
(i.e. a general purpose (digital) CMOS process).

4.1.2 Area trade-offs

The area of a VLSI circuit may be traded-off with arithmetic precision and speed requirements. These trade-offs exist at all levels of
the design hierarchy including the device level, the building block
level and the system level.

Increasing the area of a VLSI circuit component is the fundamental technique for reducing fractional circuit parameter variations
within a circuit and across many replicas of the circuit. Hence,
there is often a direct trade-off between the circuit area and the
achievable arithmetic precision. This is illustrated by considering
a differential pair transistor circuit, (see Section 3.7.2), which is
a common circuit element for multipliers, comparators and amplifiers. As the size of the transistors is increased the impact of gain
mismatch due to lithography errors is reduced and hence the input voltage offsets are reduced. Another example of device level
area trade-off, is the sizing of a storage capacitor for storing analog voltages, as may be found in some synapse designs. Here the
precision of voltage that may be stored increases with the size of
the capacitance due to the improved signal to noise ratio of the
storage device.

Precision can also be traded for area at the circuit level. For
example, a multiplying DAC circuit can be realized by generating binary weighted currents from an array of weighted current
sources or by generating binary weighted voltages or charges using
weighted arrays of resistors or capacitors respectively. In each case
the size of the device arrays determines the precision (number of
bits) that can be realized.

The relative area attributed to a particular function also requires
careful consideration at the architectural level. For example, in the
case of a fully connected recurrent neural network the number of
synapses grows quadratically with the number of neurons. Denote
the number of neurons in each layer of a m layer network by N_{L1},
N_{L1}, ..., N_{Lm} where N_{L1} is the number of linear inputs to the
network. Then for a fully interconnected recurrent network the
number of synapses in the network S_T is given by,

$$S_T = N_T^2 + N_T N_{L1} \tag{4.1}$$

where N_T is the total number of neurons and is given by,

$$N_T = \sum_m N_{Lm} \qquad (4.2)$$

For the case of a three layer MLP the total number of synapses is given by,

$$S_T = N_{L1}N_{L2} + N_{L2}N_{L3} \qquad (4.3)$$

Figure 4.1 shows the number of synapses required for a given number of neurons for neural networks of different architectures. Clearly the total area of the system can be minimised by moving circuit complexity from the synapses into the neurons. For a given network architecture and number of neurons the ratio of neurons to synapses is an important indicator of how to trade area and circuit complexity between the two building blocks.

In contrast, reducing the area of a device circuit or system is often a means of increasing the speed of operation. At the device level this is apparent as smaller devices have lower stray resistance and capacitance associated with them leading to lower RC time constants and signal propagation through the circuit. Also, the smaller the devices are, the less thermal noise they generate. Hence, increasing device size in order to improve their precision could reduce the overall system signal to noise ratio if there are many such devices.

Furthermore, the smaller the building blocks themselves are, the smaller the wiring area needed for the interconnection of the blocks and hence the less time needed for the signals to propagate through the wiring. At the system level, partitioning of functions into blocks can also have a large effect on their adjacency and hence the cost of interconnection.

4.1.3 Power trade-offs

The speed of a circuit is closely related to its power consumption. The higher the speed requirement (operations per second) the higher the power consumed. In the last few years, power consumption has become an increasingly important consideration with the advent of portable equipment and the ever decreasing device sizes available. In the former case one wishes to achieve the highest possible battery longevity. In the latter case, higher device densities implies greater heat extraction requirements, so one wishes to reduce system heat dissipation.

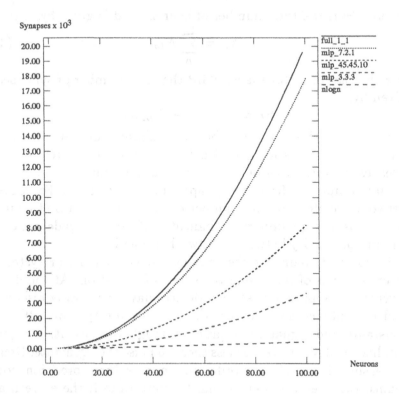

Figure 4.1 *Comparison of the rate of growth in the number of synapses as the number of neurons increases linearly. Relationships are shown for a fully interconnect network with as many inputs as neurons, a three layer MLP with a layer ratio of 7:2:1, a three layer MLP with a layer ratio of 45:45:10, a three layer MLP with a layer ratio of 3:3:3, and nlog(n).*

Assuming that a speed requirement for a computation is specified, and a circuit design for that computation exists, the question then is how to minimise the power consumption. As power is proportional to the square of voltage the obvious strategy is to reduce the power rail as much as possible. However, the consequence of doing this is to reduce the dynamic range of signals that may be represented (and hence their precision). This approach is limited by the threshold voltage of the CMOS transistors. Many analog circuits can be operated in weak inversion, however it is the upper end of the weak inversion regime which leads to the best energy time product for a circuit. Hence, two to three threshold voltages are required to operate analog circuits. Similarly, very few systems are ever purely analog due to the need for multiplexing, digital memory etc. Digital circuits require at least one threshold voltage to operate. Some researchers have examined the trade-off of lowering the threshold parameter of a process to improve power consumption [Burr and Shott (1994)].

There are also power consumption trade-offs at the circuit device level. Given a time constraint for an operation to be performed, the issue then is to minimise the energy usage of the operation. It turns out that the energy usage of a transistor varies according to its mode of operation and that there is a non-linear relationship between the energy used and the bandwidth of the circuit. The energy is a minimum and is constant over the weak inversion bias range of the transistor. Thus, the energy time product of the circuit is a minimum at the highest bias such that the transistor operates in weak inversion. However, the bandwidth of the transistor in weak inversion mode may be many orders of magnitude less than in strong inversion, thus a highly parallel architecture may be required to exploit this aspect of transistor operation (see section 3.6). This is yet another reason for considering the artificial neural network paradigm in that the most parallel circuit architecture can also be the most energy efficient. However, there are always some overheads in parallelising a computation due to increasing the fan-in and fan-out wiring and any event sequencing that may be required. This may limit the extent to which power consumption can be reduced at the expense of increased area.

4.1.4 System level trade-offs

The design of network building blocks is also influenced by a number of system level trade-offs including I/O requirements, packaging, testability and system level noise.

The computation bandwidth of a VLSI system must be matched to the available I/O bandwidth in order not to waste resources. Often the I/O bandwidth is limited by packaging requirements specifying a relatively low number of I/O pads. Furthermore, I/O requirements may vary according to the surrounding system being either a PCB with higher drive requirements than say an MCM (Multi-Chip-Module). Pads requiring higher drive can have a serious impact on power consumption. Thus, a balance must be struck between the resources required to perform a computation and the cost of communicating the inputs and outputs.

Limited output pads and the need to drive external loads introduces multiplexing and buffering requirements. In turn, multiplexers and buffers can increase the area, power consumption and reduce the precision of the signals communicated. A closely related issue is that of testability of the internal states of the VLSI circuit. In order to make an internal state observable, multiplexing is often required and some I/O overhead maybe introduced. Again the impact of the these system level requirements demands early consideration in the design cycle.

System level noise may also significantly influence the viability of a design. System level noise manifests itself in the input signals and the power supplies of the chip. Higher system noise levels can lead to higher signal and hence power rail voltage requirements and the need for differential signalling and noise cancellation circuits. These considerations can lead to increased area, power consumption and reduced precision.

4.2 Neurons and synapses

In this section we illustrate the trade-offs discussed in the previous section by way of some design examples which have been successfully implemented. Brief mention will also be given to the pitfalls encountered with these and other designs.

4.2.1 Weighted current MDAC synapse

In this subsection we describe a synapse which has been successfully used to implement a number of low power MLP chips. [Leong and Jabri (1992b); Coggins and Jabri (1994); Pickard, Jabri and Flower (1993); Coggins, Jabri, Flower and Pickard (1995)]. The synapse circuit is shown in Figure 4.2. The synapse has five bits plus sign weight storage which sets the bias to a differential pair performing two quadrant multiplication. Four quadrant multiplication is achieved by the four switches at the top of the differential pair. The bias references are derived from a binary weighted current source which distributes references to all synapses in the network. The least significant bit bias current to the synapse has been operated over the range of $100pA$ to $100nA$ and has shown good monotonicity in the weak inversion region, see Section 3.3.

The differential signalling between synapses and neurons produces good power supply noise immunity. By acting as a current sink at its outputs the synapse outputs may be tied together at the input of the neuron circuit, utilising Kirchoff's Current Law to perform the summation function. Also, the transconductance transfer characteristic (voltage input to current output) of this synapse is of the form $f(x) = tanh(x)$ (hyperbolic tangent) thus obviating the need for a non-linear squashing function in the neurons in hidden layers of a network. That is, the non-linear squashing function is distributed across the synapses of the preceding layer of the network. Thus, the functionality of the neuron required is simply that of current to voltage conversion with variable resistance (gain).

The synapse transfer function is derived as follows. The drain current of a saturated MOS transistor in weak inversion (neglecting the body effect) is given by (see Section 3.3),

$$I_{ds} = I_{do}e^{\kappa V_{gs}} \qquad (4.4)$$

where $\kappa = \frac{1}{nU_T}$. So summing currents at the tail of the differential pair we get,

$$I_{DAC} = I_o e^{\kappa(V_+ - V_-)} + I_o e^{\kappa(V_- - V_S)} = I_+ + I_- \qquad (4.5)$$

where V_S is the common source voltage of the differential pair. Now,

$$e^{-V_S} = \frac{I_{DAC}}{I_o}\frac{1}{e^{\kappa V_+} + e^{\kappa V_-}} \qquad (4.6)$$

therefore

$$I_+ = \frac{I_{DAC}e^{\kappa V_+}}{e^{\kappa V_+} + e^{\kappa V_-}} \qquad (4.7)$$

and

$$I_- = \frac{I_{DAC}e^{\kappa V_-}}{e^{\kappa V_+} + e^{\kappa V_-}} \qquad (4.8)$$

therefore

$$I_+ - I_- = I_{DAC}\frac{e^{\kappa V_+} - e^{\kappa V_-}}{e^{\kappa V_+} + e^{\kappa V_-}} \qquad (4.9)$$

$$= I_{DAC}tanh(\kappa(\frac{V_+ - V_-}{2})) \qquad (4.10)$$

This type of synapse represents a trade-off between digital and analog techniques. On the one hand, the weights are stored digitally avoiding many problems associated with the storage, reading, writing and refreshing of analog weights. In contrast the multiplier

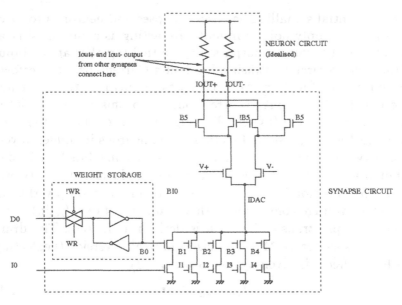

Figure 4.2 *A weighted current MDAC synapse.* © *1995 IEEE [Coggins,*
Jabri, Flower and Pickard (1995)].

is an analog circuit which is more efficient in terms of the power,
area and speed than its digital counterpart. However, it is important to realise that this particular digital storage weight scheme
is only viable for low weight precision. This is due to the area required by the binary weighted current source which produces the
current references for the synapses. The size of the current source
does not scale well with increasing number of bits. For this synapse
the area of the weighted current source (producing five references)
is approximately ten times that of the synapse. The size of the
current source doubles with each bit of precision added. Hence,
for networks of a few tens of neurons the current source rapidly
becomes prohibitively large for higher bit precisions. This synapse
is examined in more detail in Chapter 7, Section 7.2.2.

4.2.2 Current source for weighted current MDAC synapse

This current source provides the weighted current references for
the MDAC synapse. The current source (Figure 4.3) is constructed
by summing unit current sources. For transistors with uncorrelated
matching properties, summing N unit current sources improves the
matching by a factor of $\sqrt{N}$ [Bastiaansen, Wouter, Schouwenaars,

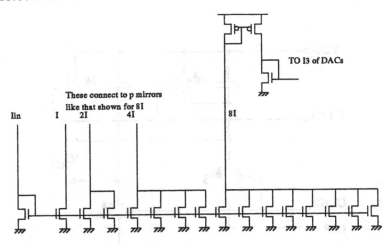

Figure 4.3 *Current source circuitry (4 bits shown). All of the n-transistors on the bottom of the circuit are the same size so $I = I_{in}$.* © *1995 IEEE, reproduced with permission.*

and Termeer (1991)]. Correlated matching properties such as variations in doping or oxide thickness are addressed by arranging the current sources in a common centroid configuration (see Section 4.3.2). Large transistors are used for the current source although smaller transistors are used inside the DACs in order to keep the total synapse area small.

The bias current is controlled by an off-chip current or voltage. Since all of the currents feeding the synapses are derived from this single input, the entire circuit can be switched off by making I_{in} equal to zero. The current source can operate in either strong inversion or weak inversion, depending on the magnitude of the bias current.

4.2.3 A switched capacitor trainable gain neuron

The neuron described in this section was designed to be used in conjunction with the synapse described in Section 4.2.1. The neuron circuit is shown in Figure 4.4. The neuron requires reset and charging clocks. The period of the charging clock determines the gain. Buffers are used to drive the neuron outputs off chip to avoid the effects of stray pad capacitances on output neurons. The reset and charging clocks must be overlapped in order that the stray ca-

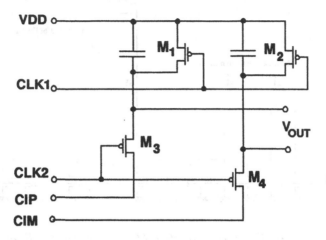

Figure 4.4 *A switched capacitor neuron.*

pacitances on the summing nodes at the inputs to the neuron are precharged high prior to the discharging phase when the output value of the neuron is computed.

The main advantages of this neuron is its flexible gain control scheme, very high equivalent resistances can be achieved and the associated ability to vary the gain in small linear increments. This last feature means that the neuron gain can be readily incorporated into gradient descent based training algorithms. Consequently, this allows the training algorithm to optimise the trade-off between magnitudes of the synaptic weights versus the value of the synapse block exponent at the neuron.

This neuron also has a number of non-idealities associated with it. One of these is a gain cross talk effect between the neurons. The mechanism for this cross talk was found to be transients induced on the current source reference lines going to all the synapses as individual neuron gains timed out. The worst case cross talk coupled to a hidden layer neuron was found to be a 20% deviation from the singularly activated value. However, the training results of the chip based on this neuron design (described in Section 10.2) do not appear to suffer significantly from this effect.

A related effect is the length of time for the precharging of the

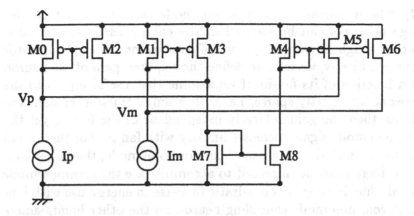

Figure 4.5 *A common mode cancelling neuron.* © *1995 IEEE [Coggins, Jabri, Flower and Pickard (1995)].*

current summation lines feeding each neuron due to the same transients being coupled onto the current source when each neuron is active. The implication of this is an increase in energy per classification for the network due to the transient decay time. A detailed case study of an implementation of the neuron is presented in Section 10.2.

4.2.4 A common mode cancelling neuron

In this section we describe an alternative neuron to that of the previous section. For larger networks with neuron fan-in much greater than ten, the problem of common mode cancellation is encountered. That is, as the fan-in increases, a larger common mode input voltage range is required or a cancellation scheme using common mode feedback is needed to overcome the effect of large common mode currents appearing at the neuron inputs.

The neuron of Figure 4.5 implements such a cancellation scheme. The mirrors M_0/M_2 and M_1/M_3 divide the input current and facilitate the sum at the drain of M_7. M_7/M_8 mirrors the sum so that it may be split into two equal currents by the mirrors formed by M_4, M_5 and M_6 which are then subtracted from the input currents. Thus, the differential voltage $V_p - V_m$ is a function of the transistor transconductances, the common mode input current and the feedback factor. The gain of the neuron can be controlled by varying the width to length ratio of the mirror transistors M_0 and

M_1. The importance of a common mode cancellation scheme for large networks can be seen when the energy usage is compared to the straight forward approach of resistive or switched capacitor neurons. Firstly, we need to define the required gain of the neuron as a function of its fan-in. If we assume that useful inputs to the network are mostly sparse, i.e. with a small fraction of non-zero values, then the gain is largely independent of the fan-in, yet the common mode signal increases linearly with fan-in. For the case of a neuron which does not cancel the common mode, the power supply voltage must be increased to accommodate the common mode signal, thus leading to a quadratic increase in energy use with fan-in. A common mode cancelling neuron on the other hand, suffers only a linear increase in energy use with fan-in since extra voltage range is not required and the increased energy use only arises due to the linear increase in the common mode current.

The transfer function of the neuron may be derived as follows. Denote the currents at the drains of M_0 and M_1 by I_{pd} and I_{md} and the mirror ratios sensing the common mode current by $\alpha = L_0/W_0 = L_1/W_1$ (since $W_2 = L_2 = W_3 = L_3 = 3.6\mu m$) and where W and L are the width and length of the transistors respectively. Summing the currents at V_p and V_m and assuming perfect matching of the transistors we obtain,

$$I_p = I_{pd} + \alpha(I_{pd} + I_{md}) \qquad (4.11)$$

$$I_m = I_{md} + \alpha(I_{pd} + I_{md}) \qquad (4.12)$$

Hence, the common mode and differential currents across the diode connected transistors M_0 and M_1 are,

$$\Delta I = I_p - I_m = I_{pd} - I_{md} \qquad (4.13)$$

$$I_{cmd} = \frac{I_{pd} + I_{md}}{2} = \frac{I_p + I_m}{2(1 + 2\alpha)} = \frac{I_{cm}}{1 + 2\alpha} \qquad (4.14)$$

whereupon we see that the differential current is preserved and the common mode current is reduced by a factor $(1 + 2\alpha)$. Using equations 4.13 and 4.14 and assuming weak inversion and strong inversion diode equations respectively, the differential output voltage is given by,

$$V_p - V_m = nU_T \log\left(\frac{I_{cm}/(1 + 2\alpha) + \Delta I/2}{I_{cm}/(1 + 2\alpha) - \Delta I/2}\right) \qquad (4.15)$$

$$V_p - V_m = \frac{1}{\sqrt{\beta}}\sqrt{I_{cm}/(1 + 2\alpha) + \Delta I/2}$$

$$-\frac{1}{\sqrt{\beta}}\sqrt{I_{cm}/(1+2\alpha)-\Delta I/2} \qquad (4.16)$$

where n is the slope factor, U_T is the thermal voltage and β is the diode equation constant determined by the transistor aspect ratio, mobility and oxide capacitance. Both of these equations require that

$$\frac{1}{(1+2\alpha)} > \frac{\Delta I}{I_{cm}} > -\frac{1}{(1+2\alpha)}$$

which corresponds to the condition for forward conduction of the diodes M_0 and M_1.

Thus, as the common mode rejection is increased by increasing α, the fractional range of the differential input around the common mode input which is linear is reduced. Using the strong inversion equation, the small signal equivalent resistance (gain) of the neuron is given by,

$$R_{eq} = \frac{1}{2\sqrt{I_{cm}/(2\alpha+1)\beta}}$$

$$= \sqrt{\frac{\alpha(2\alpha+1)}{4I_{cm}k_p}} \qquad (4.17)$$

where $\beta = \frac{W_0}{L_0}k_p = \frac{W_1}{L_1}k_p$ and substituting $\alpha = \frac{L_0}{W_0} = \frac{L_1}{W_1}$.

Further details on this neuron are given in the case studies in Chapters 7 and 8.

4.2.5 A synapse with inbuilt learning functions

In this section we describe a synapse which has hardware support for perturbation based learning algorithms. The synapse utilises a Gilbert multiplier to provide four quadrant multiplication of the differential input voltage with a weight value stored as a voltage on a capacitor. The weight is a differential value but one of the pair of signals is kept at a fixed voltage. The synapse also contains the following circuitry;

- Random access address decoding.

- A current source and sink to enable the weight value to be perturbed.

- A single bit multiplier for implementing update strategies based on approximate gradient descent algorithms.

- A single bit storage cell to hold the perturbation sense bit.
- The front end of a distributed comparator.
- Analog weight value programming.

The synapse consists of three subsections that are shown in Figure 4.6. These are:

- multiplier, weight storage and weight comparator (a),
- weight perturbation and weight write (b),
- random access addressing and decode, weight update rule and perturbation sense storage (c).

The operation of the synapses is as follows:

A differential weight voltage, presented on signals $VWEIGHT$ and VWM is applied to the differential pair biased by an nMOS-FET driven by the bias voltage $VBIAS$, (see Figure 4.6a). Another differential pair cascaded on top of each of the lower differential FETs multiplies the differential input signal, presented on signals VP and VM, with the weight voltage. This is the standard Gilbert Multiplier configuration, with the product represented by a pair of differential currents, signals IP and IM. A second differential pair, also biased by signal $VBIAS$ acts as the front end of a comparator, of signals $VWEIGHT$ and $VWPROG$, generating the differential current signal on $ICMPP$ and $ICMPM$. These signals are gated by the signal $CMPEN$ which is decoded by the random access address decoding.

The voltage on the storage capacitor is the signal $VWEIGHT$. This can be directly set to the voltage applied to $VWPROG$ by asserting both signals $RFRSHEN$ and $DORFRSH$ or it can be incrementally modified by the charge pump and sink circuitry. The weight voltage can be incremented by asserting the signal UP and biasing the pump FET with a voltage on signal $VBINC$, similarly, to decrement the weight voltage it is necessary to assert the signal $DOWN$ and bias the sink FET with a voltage on signal $VBDEC$. The size of the increment or decrement is dependent upon the duration that UP or $DOWN$ is asserted and the current that flows through the pump or sink FET due to the bias voltage. The signals UP, $DOWN$ and $DORFRSH$ are generated by the circuitry in Figure 4.6c.

The random access address decoding operates on a grid basis so that both signals RS and CS must be asserted for the synapse to be addressed. Once it is addressed then the signals $CMPEN$,

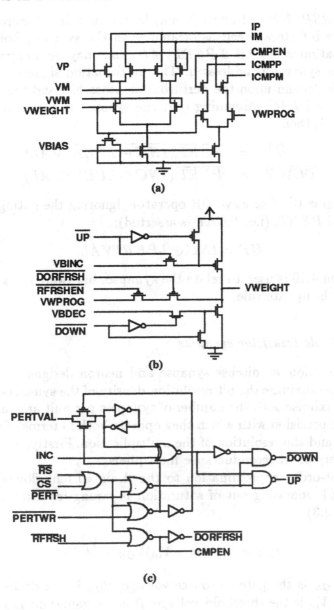

Figure 4.6 *Schematics of the three components of the synapse. (a) The multiplier, weight storage and comparator front end, (b) weight write and perturbation, (c) address decode, weight update rule and perturbation sense storage.*

$DORFRSH$, UP and $DOWN$ may be asserted, (see Figure 4.6b). A single bit storage cell associated with the synapse, holds the perturbation sense bit $LPERTVAL$, and may be programmed once the synapse is addressed. The perturbation signals UP and $DOWN$ depend upon the perturbation sense bit, and the signals $PERT$ and INC. Assuming that the synapse is currently being addressed, then;

$$\overline{UP} = \overline{PERT}.\overline{(INC \oplus LPERTVAL)} \qquad (4.18)$$

$$\overline{DOWN} = \overline{PERT}.(INC \oplus LPERTVAL) \qquad (4.19)$$

Where $\oplus$ is the Exclusive OR operator. Ignoring the gating effect of signal $PERT$, (i.e. $PERT$ is asserted);

$$UP = INC \oplus LPERTVAL \qquad (4.20)$$

Equation 4.20 is used to relate the synapses hardware design with the weight update rule.

4.2.6 Single transistor synapses

In this section we discuss synapse and neuron designs which attempt to maximise the bit resolution density of the synapses. That is, to maximise both the number of synapses per unit area and the effective precision with which they operate both in terms of weight storage and the resolution of the multiplication. Firstly, we review the theory of transconductance multipliers.

A first-order approximation to the drain current flowing in a MOSFET operating out of saturation in strong inversion is, (see Section 3.3)

$$I_{DS} = \beta\{(V_{GS} - V_{t0})V_{DS} - \frac{V_{DS}^2}{2}\} \qquad (4.21)$$

where V_{GS} is the gate to source voltage, V_{DS} is the drain-source voltage, V_{t0} is the threshold voltage, β is the gain, and I_{DS} is the drain current and the slope factor n has been assumed close to one.

If two identical transistors are then configured as in Figure 4.7, where $V_{DS1} = V_{DS2}$ then the difference of the two drain currents is calculated via equation 4.22, which is the scaled product of V_{DS} and $(V_{GS1} - V_{GS2})$. Four quadrant multiplication is achieved as equation 4.21 still holds when V_{DS} is negative but with the definition of the source and drain interchanged.

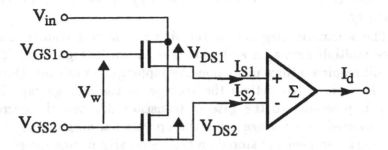

Figure 4.7 *Transconductance multiplier scheme.*

$$I_d = I_{DS1} - I_{DS2} = \beta V_{DS}(V_{GS1} - V_{GS2}) \qquad (4.22)$$

The output of this multiplier is a current, therefore, a vector product may be performed by simply connecting the sources of corresponding transistors together.

$$\Sigma I_d = \beta \Sigma V_{in} V_w \qquad (4.23)$$

where $V_{in} = V_{DS}$ and $V_W = (V_{GS1} - V_{GS2})$.

Two difficulties exist in the circuit of Figure 4.7. The first being the need to ensure that the drain to source voltages across both transistors are equal. This can be achieved by connecting the sources to a virtual ground. The second is that the two transistors may have some systematic differences which will manifest themselves in error terms that will be dependent upon the difference of the gains and threshold, such that;

$$I_a = I_d + (I_{EVT} + I_{E\beta}) \qquad (4.24)$$

where I_a is the actual current and,

$$I_{EVT} = \beta(\Delta V_T V_{DS}), I_{E\beta} = \Delta\beta\{(V_{GS} - V_T)(V_{DS} - \frac{V_{DS}^2}{2})\} \quad (4.25)$$

are the threshold and gain terms respectively. This second problem can be alleviated by careful matching of the transistors or alternatively using a single transistor and switched capacitor multiplexing. This means that currents I_{S1} and I_{S2} are derived from the same transistor and the values of $\Delta\beta$ and ΔV_T are zero. The error terms are effectively removed from the multiplication term, however, there is some error introduced by the switched capacitor circuitry.

The schematic diagram for the single transistor transconductance multiplier synapse and neuron is shown in Figure 4.8. This circuit requires a three phase non-overlapping clock scheme. During the first phase (ϕ_1 active) the voltage on the storage capacitor C_{wh} is presented to the gate of transistor M_1 and the current is converted to a voltage using the operational amplifier A. This voltage is sampled and stored on $C_{\phi1}$. Similarly, during the second phase (ϕ_2 active) the voltage on C_{wl} is presented to the gate of M_1 and the resultant voltage is stored on $C_{\phi2}$. During the third phase (ϕ_3 active) the difference of $C_{\phi1}$ and $C_{\phi2}$ is generated and stored in C_S. The output voltage is then proportional to the sum of the product of the input voltages and the differential weight voltages, such that;

$$V_o = (\beta \sum_m V_{in_m}(V_{wh_m} - V_{wl_m})) + V_E \qquad (4.26)$$

where V_E is the error introduced by the storage capacitor non-linearities. Note, since the product is calculated as the difference of two voltages generated in the same operational amplifier any offset in the amplifier will be cancelled.

The synaptic density of the single transistor design STTM is high, 320 synapses/mm^2 allowing a large die (1 cm^2) to contain 25,000 to 30,000 synapses. The equivalent bit precision of the multiplier is very high, being an analog multiplier, however the synaptic bit precision is limited by the weight storage element which is a capacitor. Previous work [Pickard, Jabri, Leong, Flower and Henderson (1992)] with these synapses and [Jabri, Pickard, Leong, Rigby, Jiang, Flower and Henderson (1991)] suggests that 10 to 12 bits of precision can be achieved. The neuron to synapse size ratio is 10.35, which indicates that for networks of greater than 10 neurons (i.e. 100 synapses) the total size of the neurons is less significant than the total size of the synapses. The synapse and their multipliers are relatively easy to design, however, the neurons present a somewhat more challenging task as they must provide a virtual earth, perform current to voltage conversion, and output a common mode voltage compatible with the next input stage. These techniques can potentially produce the best bit resolution density available for analog ANNs.

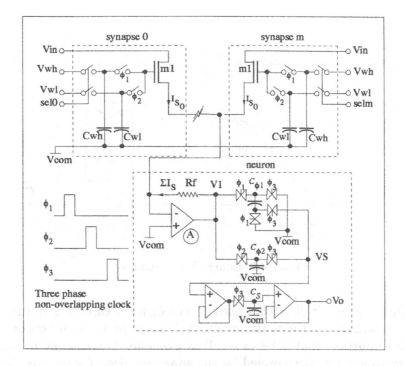

Figure 4.8 *Single transistor synapse and neuron.*

4.2.7 A charge based synapse

The essential functions of a feed forward network are multiplication, summation and a non-linearity. Summation is commonly performed by simply summing currents at a node. The alternative used here is to sum charge. The advantage of this is that charge is simply the product of voltage and capacitance and so the multiplying element design is straight forward. [Gregorian, Martin and Temes (1983)] reviews the design principles of charge amplifiers and capacitor arrays.

The synapse consists of a weight storage element and a multiplier as illustrated in Figure 4.9. As charge is the desired resultant product, multiplication is the product of voltage and capacitance. Hence, a variable weight is realised by having a variable capacitance. Four quadrant operation is realised by using a crossover switch to control the connection orientation of the synapse capacitor to the summing node and connecting the positive input of the charge amplifier to a suitable reference voltage. The weight store

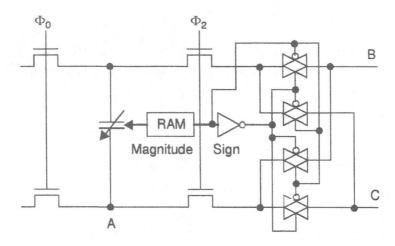

Figure 4.9 *Charge based synapse.*

is just a number of latches used to configure a capacitor matrix. As the synapse is the most numerous element in a neural network, it is important that it be as small as possible. The minimum unit capacitance was determined by the space required for routing the connection grid, and this turned out to be of significance, see Section 10.1.1. Size, rather than matching or power consumption, is therefore the limit on resolution. The clocks Φ_0 and Φ_2 are used to control charging and transfer of charge to the summing node respectively. From Figure 4.9 we see that single FETs are used as pass transistors rather than a transmission gate. This is because it was desired (for convenience) to have the non-linearity in the synapse rather than in the neuron. A detailed case study of an implementation of this synapse is presented in Section 10.1.1.

4.2.8 A charge based neuron

This is merely a charge amplifier with a reset switch across the feedback capacitor as is shown in Figure 4.10. The labels B and C correspond to those in Figure 4.9 previously.

The magnitude of the feedback capacitor is controlled by a series of latches that configure a matrix, as in the synapse. Four bit gain resolution (no sign bit) was considered adequate at this stage. The reference voltage V_r permits four quadrant operation as the output can swing above or below this value. The value for the bias current

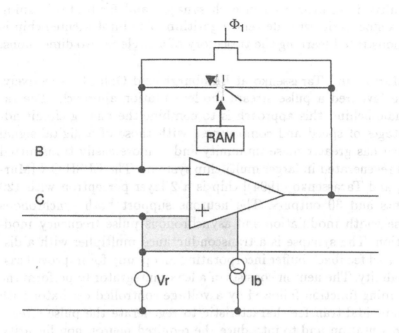

Figure 4.10 *Charge based neuron.*

is a function of the clock speed. One has to be able to charge
the weight capacitors of the following stage in the time available.
The lower the bias current the less output current is available for
charging. A detailed case study of an implementation of this type
of neuron is presented in Section 10.1.2.

4.2.9 Other implementations

For completeness we mention here the work of other researchers
who have had some success in analog VLSI neural network implementations with learning capability. For detailed discussions, the
reader is referred to the references provided.

Cauwenberghs has demonstrated a continuous time analog VLSI
recurrent neural network design with six fully interconnected recurrent neurons (see Section 8.3). The neurons and synapses are constructed from CMOS wide range transconductance elements which
are used to form a weight multiplier in the synapse and a floating
resistor in the neuron which, along with a parallel capacitor, determine the continuous time neuron dynamics. Perturbation based

circuitry is incorporated in each synapse and facilitates learning
via a stochastic error descent algorithm. The implemented chip is
demonstrated learning the trajectory of a circle in two dimensions.

Murray and Tarassenko at Edinburgh and Oxford respectively,
have favoured a pulse stream implementation approach. The ra-
tionale behind this approach is to combine the analog circuit advantages of speed and compactness with those of a digital signal
which has greater noise immunity and is more easily transmitted
and regenerated in larger multi-chip systems. The EPSILON [Mur-
ray, and Tarassenko (1994)] chip is a 2 layer perceptron with 120
inputs and 30 outputs. The neurons support both synchronous
pulse width modulation and asynchronous pulse frequency modulation. The synapse is a transconductance multiplier with a distributed feedback buffer incorporating an op amp for improved cas-
cadability. The neuron consists of a leaky integrator to perform the
summing function followed by a voltage controlled oscillator with
a sigmoidal transfer characteristic to regenerate the pulse stream
representation and to introduce the required neuron non-linearity.
Weights are stored capacitively in the synapses and are periodically
refreshed by off-chip RAM. The chip is demonstrated to perform
a vowel classification problem in an MLP architecture.

4.2.10 Summary

The preceding sections are summarised in Table 4.1 which shows
the type of circuit, its function, the mask layout area and a com-
parative power rating.

4.3 Layout strategies

In this section we discuss how the building block circuits of the previous section may be physically mapped to silicon for fabrication.
Since any given cell is intended to form an integral part of a larger
system (layout) both the techniques that apply intra cell and at
the chip level are discussed. For example, intra cell issues include
layout for best component matching and techniques for minimizing
cell area while chip level issues include designing cells for abutment,
minimizing routing complexity and minimizing wiring costs.

Table 4.1 *Summary of the circuits described in Section 4.2 listing the design technique, area and relative power requirements of each of the circuits.*

Circuit	Design Technique	Area (μm^2)	Power
MDAC Synapse	Digital Analog	23.7×10^3	low medium
Switched Capacitor Neuron	Switch Capacitor	32.0×10^3	low
Common Mode Cancelling Neuron	Analog	8.3×10^3	low medium
Inbuilt Learning Synapse	Digital Analog	23.7×10^3	low medium
Single Transistor Synapse	Analog	4.0×10^3	medium high
Charge Based Synapse	Analog	42.1×10^3	low
Charge Based Neuron	Analog	25.6×10^3	low

4.3.1 Individual cell design

The layout of a building block can significantly affect the circuit performance. Some of the performance parameters which can be affected by layout include:

- Noise injection from power lines, clock lines and the substrate.
- Clock feed through.
- Component matching.
- High frequency response of the circuits.
- Linearity.
- Sensitivity to process variations.

Since component matching is often a very important aspect of a design, especially for circuits designed to operate in weak inversion mode, the discussion of these techniques is presented separately

in the following subsection. Component matching aside, the afore-
mentioned performance aspects of the circuits may be improved by
minimizing the layout parasitics. Layout parasitic effects include
capacitive coupling, series resistance, leakage currents and long distance coupling either capacitively through air or resistively through
the substrate. These effects can be overcome by first identifying the
critically effected nodes and loops in a circuit and minimizing the
parasitics or compensating for them with dummy structures.

Capacitive coupling can be reduced by using minimum sized devices and the minimum number of devices (as in the case of minimizing clock feed through by analog switches) and then spacing
critical nodes as far apart as possible. Grounded metal lines and
polysilicon layers may also be used to shield nodes from each other
and from the substrate. The substrate may also be shielded by a
separately grounded well.

Series resistance is important in the layout of power supply lines.
This is especially the case in mixed analog and digital circuits.
In this case, a common power bus can mean that digital broad
band signals are coupled to the analog supply. This problem can
be overcome by making the buses of power supplies wide, separating analog and digital supply lines and, better still, providing
the supplies from different external pins where they can also be
separately bypassed.

Resistive coupling through the substrate can inject noise from
one circuit into another. Firstly, one should minimize the noise coupled into the substrate by using the shielding techniques already
mentioned and also using special wells to isolate or shield circuits.
Given this, the designer should distance circuits from points of
noise injection as far as possible and make critical circuits as compact as possible to minimize the effect of substrate voltage gradients. Coupling of minority carriers from switching circuits also occurs through the substrate from blocked transistor channels. This
can be overcome by separating wells and using wells as shields. For
further reading on these issues see [Gregorian and Temes (1986)].

4.3.2 Layout for better component matching

Component matching is often critical to circuit performance. For
example, a differential pair amplifier depends on good matching of
the transistors to achieve low input offset and high power supply
rejection ratio. Similarly, matching of capacitors and resistors in

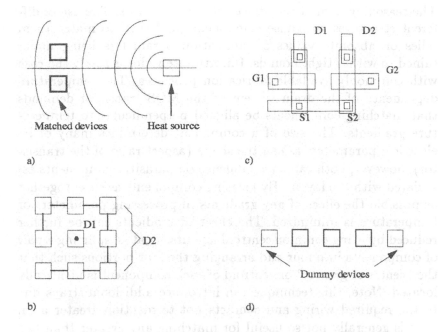

Figure 4.11 *Examples of various layout techniques to overcome component matching difficulties in CMOS VLSI circuits. a) Matching for same temperature. b) Common centroid geometry. c) Matching for same orientation. d) Matching with same surroundings.*

array configurations are important for achieving monotonicity and linearity for digital to analog converters. The following rules should be followed to achieve the best possible component matching, some of which are illustrated in Figure 4.11:

- Components realised using the same structure. e.g. a gate capacitance will not match well to a double polysilicon capacitor.

- Locate components in regions of equal temperature.

- Layout components with the same shape and equal size.

- Locate components as close to each other as possible.

- Use common centroid geometries.

- Place components in the same orientation.

- Size components above minimum size.

- Provide the same surroundings for components.

The reasoning behind the above rules is as follows. The use of different structures to realise components and then to match them relies on absolute values of fabrication parameters being maintained to within tight bounds. Unfortunately this is rarely the case with commonly available fabrication processes. The temperature dependence of the drain current of the MOS transistor demands that matching components be aligned perpendicular to temperature gradients. The size of a component determines many of its electrical parameters as can the shape (aspect ratio of the transistor), however, both can have influence on parasitic components associated with the layout. By locating components as close together as possible the effect of any gradients in processing parameters or temperature is minimized. The effect of gradients can be further reduced by using common centroid layouts. That is, splitting a pair of components into four and arranging the four portions such that the "center of gravity" or centroid of each component is identically located. Note, this technique can introduce additional strays due to the required wiring and contacts not to mention greater area, so it is generally not so useful for matching any greater than two devices. Placing components in the same orientation is another simple way of reducing the impact of gradients. By increasing the size of components above minimum size, spatial fluctuations in process parameters and fluctuations in component dimensions are averaged out. Surrounding structures may effect a component by parasitic coupling and leakage effects. Where the effects of surrounds are understood, dummy structures could be used to improve matching. For further discussion see [Vittoz (1991)].

With the above rules in mind one has to adapt them for the case where non-unity ratios of component values are required. In the case where a ratio of integral values is required one can break the problem up into arrays of unit devices where the unity ratio matching rules apply. In the case of non-integral ratios one simply diverges from the matching rules as little as possible.

4.3.3 Floor planning, routing and I/O

In this section we discuss the layout considerations necessary for combining building blocks into a working VLSI system. It is important to realise that the layout issues in this section impact directly on the layout of individual cells, so it is necessary to maintain system assembly requirements while laying out individual cells. The layout issues involved in the assembly of a VLSI system include:

- Wiring by abutment versus free routing.
- Positioning of ports on cells.
- Choosing cell aspect ratios.
- Choice of wiring layers.
- Chip floor planning.
- Power Distribution and I/O.

Wiring by abutment is perhaps the best method for interconnecting functional blocks in that it is systematic and predictable in terms of the strays added to the circuit and the area used by a given design. However, abutment is only efficient when there are many common signals to be propagated to lines and arrays of cells, otherwise excessive wiring redundancy can result. Thus, less commonly used signals should be free routed to external ports of a block. An example of this type of choice occurs with the weighted current synapse. Initially one might consider abutting synapse arrays for MLPs so as to propagate the neuron activations. However, this is impractical since the five current references and synapse addressing outweigh the actual neuron activations in wiring cost and the aspect ratios of the synapse arrays makes for an awkward floor plan if neuron activations are routed by abutment.

Choosing cell aspect ratios, port locations and use of wiring layers are all closely linked to the chip floor plan, but are also of course constrained by individual cell layout requirements discussed in previous sections. (In regard to minimizing the impact of strays on critical nodes and loops in the circuit). Thus, estimates of the floor plan and individual cells must be made before layout begins and these estimates updated as the constraints on the layout become more certain. As far as analog circuits are concerned, sensitive parts should be well separated from switches and any digital circuitry and be shielded by wells and metal lines if necessary. The floor plan must also take into account the amount of free routing and allow space for routing channels commensurate with the sophistication of the auto router used.

Since power distribution and I/O connections are vital to good performance in many instances, it is often prudent to hand route these signals. In the case of power connections this enables the use of appropriately wide buses and the separation of noisy supplies from quiet ones. In the case of I/O, designs are often pad limited and signal lines must sometimes be shared. Care must be taken when using shared signals to ensure that no untoward coupling of supply noise occurs.

4.3.4 Electrostatic discharge and latch-up protection

It is essential to protect all gates connected to pads from electrostatic discharges caused by charge build up external to the chip. This usually consists of a scheme using diodes to clamp the gates between the power rails and large area resistors to dissipate energy. CMOS circuits must also be protected from latch-up. Latch-up is caused by the parasitic lateral pnp and npn bipolar transistors associated with the p and n diffusions of the MOS transistors. The collectors of each parasitic bipolar feed the base of the other leading to a DC current loop through the substrate. Large substrate currents will disrupt the operation of circuit elements and possibly destroy the chip. The likelihood of latch-up can be minimized by reducing the amount of current flowing in the loop. This can be achieved by having a sufficient number of well contacts and appropriate spacing of devices to minimize the DC loop gain.

4.3.5 Netlist comparision

A final point worth mentioning is netlist comparison of the mask layout and a schematic diagram of the system. This is the only way that a designer can be sure that the schematic is a true representation of the circuit implemented in VLSI. Conversely, it provides a confidence level for guaranteeing that the mask layout is a faithful execution of the schematic. Many CAD systems have a means of netlist comparison built-in but in some cases it may be necessary to extract the netlists from the schematic diagram and the mask layout and compare them using a tool such as GEMINI, NETCMP or WOMBAT. Let it suffice to say that an automated means of verifying a mask layout goes a long way towards building confidence in the integrity of a design.

4.4 Simulation strategies

4.4.1 Objectives of simulation

Before beginning a simulation exercise it is important to consider
carefully what you hope to learn from the simulation and what
impact this extra knowledge will have on the design process. The
scope of a simulation can extend from the network or system level
down to a small analog subcircuit. Ideally, a simulation environment would simulate all desired characteristics of a system down
to the smallest level of detail such that a designer could access
any node in the system at any time. Unfortunately, now and for
some time to come this is a computationally infeasible task for
large scale analog circuits such as neural networks given current
hardware platforms and software environments. Hence, a stratified
approach must be taken to break down the simulation task into
computationally feasible chunks. This either means that the circuit itself can only be simulated in smaller isolated blocks leading
to a loss of information about inter-block interactions or aggre-
gated block simulations are performed where detailed operation of
blocks is averaged or simplified.

In order to choose the most appropriate mix of these two approaches the designer needs to consider the following issues:

- simulation software available and its performance on the available hardware platform.

- the combinatorial complexity of the simulation task. e.g. the
  simulation of all possible combinations of transistor parameter
  mismatch.

- the effect the knowledge gained will have on the design process.
  (e.g., it may be prudent to simply design more conservatively
  when both the cost of simulation and likelihood of circuit failure
  are high.)

In the next sections we discuss some different simulation mixes, the
type of information they provide and their feasibility and availabil-
ity.

4.4.2 Analog cell simulation

Here we consider the simulation of a single cell or a portion of a
cell. The size of the cell would normally be a few tens to a few

hundred components. (Transistors, capacitors, resistors etc.) The objectives of simulating this type of cell are usually:

- Functional verification. i.e. does the circuit behave according to the analytical model we had in mind.

- Post layout verification. When the circuit has been laid out we can include stray capacitance and resistance due to wiring in the simulation. This answers the question as to whether the circuit is tolerant to strays, which can be important for some switched capacitor techniques for example.

- Sensitivity analysis. Some transistor parameters, for example, the threshold, can vary widely on and between chips. A functional simulation can be performed repeatedly with component parameters varied according to some statistical model (hopefully based on the fabrication process if this is known) and the circuit response checked to make sure it remains within acceptable tolerances.

The first two of these tasks can usually be performed with fairly modest resources, for example a PC running SPICE. However, be aware that some simulation models may not be adequate for the type of analysis you wish to perform. For example, it is well known that the standard MOSFET SPICE model (levels 1-4) suffers model discontinuities when a transition from the weak to the strong inversion mode is made. The EKV model [Enz, Krummenacher and Vittoz (1994)] is a model which avoids this problem. Sensitivity analysis often needs more powerful computing platforms in order to exhaustively vary device parameters. Some times a compromise may be struck if it can be determined that only the extreme limits of parameters are important.

4.4.3 Analog simulation with macro models

Once the performance of individual cells and sub-circuits has been verified for functionality, layout effects and fabrication variations it is necessary to check that performance is not degraded when the cell is integrated into a larger circuit. Since replicating a cell many times is likely to produce a computationally infeasible simulation task it is necessary to introduce some less detailed model of each cell. This might be achieved by describing the transfer function of the cell by some equations or alternatively using a look up table derived from the cell level simulations. Simplifying the simulation

model of the cell can often be done with little loss of accuracy
if macro model boundaries can be placed where one circuit block
is "buffered" or electrically isolated from a neighbouring block.
This type of simulation strategy is often useful for identifying the
following in a design:

- When cells are loaded by neighbouring circuits in the system,
  do they still have the requisite performance?

- When system level wire routing is back annotated from the layout, does the stray capacitance or resistance degrade the performance?

4.4.4 Analog simulation of small networks

Where it is impossible to decouple cells into simplified macro models, sometimes one has no choice but to attempt to simulate small
groups of cells in accordance with the computational resources
available. This approach may also be worthwhile to gain confidence
that a macromodel is adequate before using it in a simulation of
the whole system. In the case of a neural network, it is often possible to simulate a small network of a few neurons and synapses in
this manner.

4.4.5 High level functional modelling

The last form of simulation we consider is the system level simulation independent of any particular hardware implementation. That
is, building blocks of the system are modelled in a programming
language by their transfer function. This is not to say that the
effects of hardware implementation are ignored, but rather, it is
often important to model known hardware effects or assume some
hardware non-idealities. Doing this early in the design process, before any building block implementations are conceived, is useful in
demonstrating the viability and tolerance of the system level con-
cepts to analog implementation. For example, analog weight storage versus digital weight storage in a network and the associated
effective precisions of both storage mechanisms can have significant
implications for the learning and operational performance of a network. In the digital case quantization errors are introduced to the
weights, where as in the analog case the weight value has a probability distribution associated with it according to the noise model

of the weight storage mechanism. These types of implementation effects, as well as others such as offsets, other noise sources and drift can all be modelled in a system level model. The objectives of this type of modelling include:

- To verify the adequacy of a given network architecture for the problem to be solved.
- To assess the minimum arithmetic precision required for the application.
- To generate building block minimum design specifications with regard to noise performance, offsets, drift, quantization errors, power and speed.
- To provide a realistic environment for testing hardware oriented learning algorithms.

Although we have discussed this simulation technique last in that we have emphasised in this chapter the design issues for the low level building blocks, it can be seen from the above discussion that it is often necessary to attack the network design problem from both ends. A simultaneous top-down and bottom-up approach leads to a shorter design time with better results.

Taking up the theme of weight storage again, as an example, the effective weight precision determined by system level simulations can have profound effects on the weight storage techniques which are feasible and even the type of network architecture which should be used. A case in point is a requirement of high weight precision, e.g. ten bits. This may rule out a current based MDAC synapse design due to the prohibitive size of the binary weighted current source. However, this weight precision requirement may be reduced by either increasing functionality of the neuron by providing high precision gain (many bits of blockwise exponent) or increasing the number of lower precision synapses available. These types of design variants then have to be checked for other effects they introduce such as offsets or additional noise generation to establish the exact trade-offs involved. Thus, it is apparent that the designer needs to gain intuition about the interaction of system level choices and building block circuit design choices.

CHAPTER 5

Kakadu – a micropower neural network

Co-author: **Philip H. W. Leong**

This chapter introduces a basic analog VLSI neural network called 'Kakadu' [Leong and Jabri (1992a); Leong and Jabri (1993a)]. Kakadu is a 10:6:4 multi-layer perceptron with neurons implemented off chip. It is a slightly larger version of an earlier test chip [Leong and Jabri (1992b)] which contained the building block circuits of synapses, current sources, bucket brigade and a smaller 3:3:1 multilayer perceptron. The chapter describes the design criteria for building a basic network and discusses the important performance characteristics of the network building blocks to ensure useful operation. The chapter concludes by describing the performance of Kakadu on a cardiac arrhythmia classification problem.

5.1 Advantages of analog implementation

The Kakadu chip was designed to act as a classifier in a low power biomedical system, specifically, to identify abnormal morphologies of patients being treated using an implantable cardioverter defibrillator (ICD).

The Kakadu chip was implemented using low power analog techniques because they have the following advantages

- weak inversion analog circuits enable implementations which consume very little energy

- they are easily interfaced to the analog signals (in contrast to digital systems which require analog to digital conversion)

- the problem of designing analog circuits which operate over a wide temperature range is removed for implantable systems since the human body is very well temperature regulated

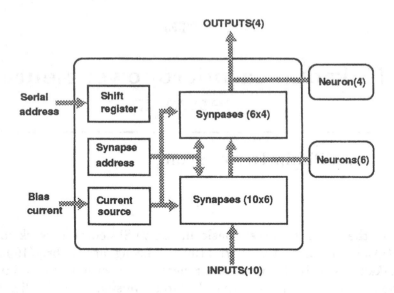

Figure 5.1 *Block diagram of the Kakadu MLP chip.*

- analog circuits are generally small in area
- fully parallel implementations are possible
- a certain amount of fault tolerance may be exhibited by the neural network.

All of these features are desirable in implantable biomedical devices.

5.2 Architecture

Kakadu implements an artificial neural network based on the multi-layer perceptron model introduced in Section 2.5. A block diagram of the chip is shown in Figure 5.1. The chip takes voltage inputs and passes them through the first array of synapses to produce six pairs of hidden layer currents. These currents are converted to voltages using linear resistors that are external to the chip. The same nodes are used as voltage inputs to the next layer which produces output currents which are converted to voltage outputs by the output neuron layer.

The main blocks of the chip are two synapse arrays, a current source and weight addressing circuitry. The synapse's digital to

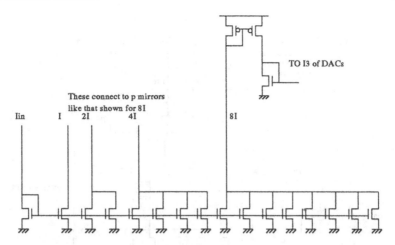

Figure 5.2 *Current source circuitry (4 bits shown). Typical bias currents used are 6.63 nA, and all of the n-transistors on the bottom of the circuit are the same size so* $I = I_{in}$. © *1995 IEEE, reproduced with permission.*

analog converters are binary weighted current sources which are controlled by digitally stored weights. A common current source is used to supply bias voltages to the DAC in each synapse. The circuit can be operated over a wide range of bias currents.

Although inputs to the neural network are analog, synapse values are written digitally. This enables configuration of the chip to be performed digitally but keeps the actual signal processing in the analog domain. The synapse array appears as an 84 word RAM (the first layer having 10×6 words and the second layer having 6×4 words) with a 6 bit word size. All biases for the chip are derived from a single off-chip pin which sets the current source bias I_{in} (see Figure 5.3.1). Synapses are addressed by row and column through pairs of multiplexed row and column shift registers.

5.3 Implementation

5.3.1 Current source

A single current source is used to provide biases for all synapses of the chip. The current source (Figure 5.2) is constructed by summing unit current sources and was previously described in section 4.2.2.

The current source least significant bit current can be varied

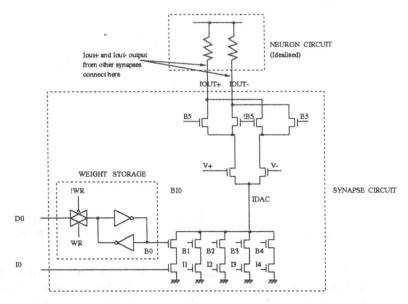

Figure 5.3 *Synapse and neuron circuitry.* © *1995 IEEE [Coggins, Jabri, Flower and Pickard (1995)].*

from strong down to weak inversion. By removing the bias altogether the synapse array may be duty cycled to save power.

5.3.2 Synapse

The synapse was described in Section 4.2.1. It is composed of registers which store the weight values, a linear DAC and a transconductance multiplier. The bias current is the same as the unit current for the DAC so each DAC can output ±31 times the bias current. The circuit diagram of the synapse is shown again in Figure 5.3.

Since synapses are the most numerous elements in a neural network, the size of the network that will fit in a given area is controlled by their dimensions. Although small synapses are required, the matching of crucial transistors (the 5 mirror transistors connected to I0–I4) within the synapse is proportional to the square root of the transistor area and so these transistors should be made as large as possible. A compromise was reached in selecting $81\mu m^2$ transistors for the I0 to I4 mirrors within the synapse.

Storage of the synapse values is achieved using registers, the value of which are converted to analog values via the DAC. To achieve a small synapse area, the registers were designed to be as narrow as possible since each register contains 6 flip–flops.

The DAC is constructed through current summing. Each bit of the DAC is controlled by a pass transistor which can be turned on or off depending on the value stored in the (static) input flip–flop (B0–B4). I0–I4 are voltages taken from the current source which serve to bias the currents in powers of two. The entire synapse array appears as a large (write only) register to the controlling digital circuitry which programs the weight values.

The DAC is connected to a transconductance multiplier to form a synapse. The multiplier has a pair of voltage inputs, a current input (from the DAC) and a pair of current outputs. As derived in section 4.2.1, the transfer function of this multiplier is given by the relation

$$I_{out+} - I_{out-} = \begin{cases} +I_{DAC} \tanh(\frac{\kappa(V_+ - V_-)}{2}) & \text{if B5} = 1 \\ -I_{DAC} \tanh(\frac{\kappa(V_+ - V_-)}{2}) & \text{if B5} = 0 \end{cases} \quad (5.1)$$

The multiplier is linear with the current inputs (from the DAC) and nonlinear to the neuron voltage inputs. This is the desired situation as if they were reversed, the tanh function would only serve to compress the range of weight values available and would not allow nonlinear problems to be solved. The DAC only produces positive values. Current switching logic controlled by B5 enables the output to be changed in sign if a negative weight is desired. The $V+$ and $V-$ inputs are from either neurons or input pins. Output of the multiplier are two current sinks.

The area of a synapse is $106 \times 113\mu m$ which includes all of the weight storage registers and switches, I0–I4 current mirrors, multiplier and sign switching circuitry. A neural network can be constructed from a single current source (described in Section 4.2.2) and a synapse array.

5.3.3 Neurons

In a low power system, where the neuron input current can be of the order of ten nA, a high impedance of the order of 1 MΩ is required. This is hard to implement in standard MOS technology because diffusion and polysilicon do not have the high resistance necessary, and an active circuit with the desired transfer characteristic is hard to design. If on–chip neurons are used, a method of measuring the activation of at least the output neurons is required for training, and this requires buffers to drive the signals off–chip.

A possible solution to this problem is to implement the neurons using off–chip resistors. The resistors are provided off chip in order to allow easy control of the impedance and transfer characteristics of the neuron. These neurons also serve as test points for the chip. It is envisaged that these neurons will later be integrated onto the chip using either an active circuit or a high resistance material such as that used in a static RAM process. Since the neurons are implemented externally to the chip, nonlinear neurons can also be used.

However, a resistor has a linear characteristic which, at first glance, appears unsuitable. This problem was addressed by implementing the nonlinear characteristic required by the neural network in the synapse instead of the neuron. Using this technique, the nonlinearity of the Gilbert multiplier is used to an advantage.

The transfer function of the Kakadu network is

$$u_i = \sum_{j=1}^{N_l} w_{ij} f_l(a_j) \tag{5.2}$$

$$a_i = \alpha u_i \tag{5.3}$$

where u_i is the summed output of the synapses, a_i is the neuron output, κ and α are constants, l denotes the lth layer ($0 \leq l \leq L-1$), L is the total number of layers (namely 2), N_l is the number of neuron units at the lth level i is the neuron number ($1 \leq i \leq N_l$) and $f_l(x) = \tanh(\frac{\kappa(x)}{2})$.

For a two layer network, Equation 5.3 is very similar to the typical multilayer perceptron model as illustrated in Figure 5.4. Any set of inputs can be scaled so that they are within the linear range of $f_l(x)$ and so the initial nonlinearity applied by $f_0(x)$ (i.e., $f_l(x)$ where $l = 0$) does not change the computational capability of the circuit. There is an absence of a nonlinearity in the final layer, and this can be thought of as a linear output unit. Equation 5.3 can thus be rewritten in the familiar multilayer perceptron form

$$u_i = \sum_{j=1}^{N_l} w_{ij} a_j \tag{5.4}$$

$$a_i = g_l(u_i) \tag{5.5}$$

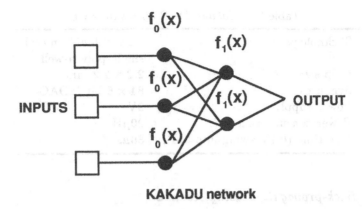

KAKADU network

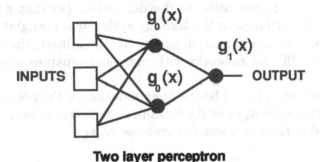

Two layer perceptron

Figure 5.4 *Comparison between the Kakadu architecture and a multi-layer perceptron. The nonlinearity is depicted by the circles and lines represent synapses.*

where $g_0(x) = \alpha \tanh(\frac{\kappa(x)}{2})$, $g_1(x) = \alpha x$ and it is assumed that the inputs have been initially passed through $g_0(x)$.

As shown in Chapter 6 this does not affect the neural network's ability to solve highly nonlinear problems such as the 4-bit parity problems. The disadvantages of using off–chip neurons are that the currents must travel through pins so pin leakage may affect the circuit and also, for larger networks, the number of pins required may become excessive. It should also be noted that all analogue VLSI neural network implementations have limited input and output ranges since they are (at best) bound by voltage and current restrictions imposed by the supplies.

Table 5.1 *Kakadu MLP chip summary*

Technology	$1.2\mu m$ double metal, single poly n-well
Chip size	2.2×2.2 mm
Synapses	84×6 bit MDAC
Power supply	3V
Power consumption	$20\mu W$
Rise time (0.2V swing into 2pF)	$30\mu s$

5.3.4 Back-propagation using Kakadu

Although the Kakadu chip is mostly trained using techniques which do not require computation of the derivatives (so that nonideal effects can be included in the training cycle), it is straightforward to derive the back-propagation equations [Rumelhart, Hinton and Williams (1986)] for networks with nonlinear synapses and linear neurons.

For a pattern p, let t_{pj} be the desired output of the network and o_{pj} be the actual output of the Kakadu style neural network. Then we can define the mean squared error as being

$$E_p = \frac{1}{2}\sum_j(t_{pj} - o_{pj})^2 \qquad (5.6)$$

If we rewrite Equation 5.3 using the same notation as Rumelhart, Hinton and Williams (1986), we get

$$\text{net}_{pj} = \sum_i w_{ij}f(o_{pi}) \qquad (5.7)$$

$$o_{pj} = \text{net}_{pj}. \qquad (5.8)$$

Back-propagation involves finding

$$\frac{\partial E_p}{\partial w_{ij}} = \frac{\partial E_p}{\partial \text{net}_{pj}}\frac{\partial \text{net}_{pj}}{\partial w_{ij}} \qquad (5.9)$$

Since $\partial \text{net}_{pj}/\partial w_{ij} = f(O_{pi})$ because $\partial w_{kj}/\partial w_{ij} = 0$ except when $k = i$, Equation 5.9 can be expanded to

$$-\frac{\partial E_p}{\partial w_{ij}} = \delta_{pj}f(O_{pi}) \qquad (5.10)$$

where $\delta_{pj} = -\partial E_p/\partial \text{net}_{pj}$.

To evaluate δ_{pj} for an output unit, using Equations 5.6 and 5.8, we see that

$$
\begin{aligned}
\delta_{pj} &= -\frac{\partial E_p}{\mathrm{net}_{pj}} \\
&= -\frac{\partial E_p}{\partial O_{pj}}\frac{\partial O_{pj}}{\partial \mathrm{net}_{pj}} \\
&= (t_{pj} - O_{pj})k \\
&= k(t_{pj} - O_{pj}).
\end{aligned}
\tag{5.11}
$$

For a hidden unit

$$
\begin{aligned}
\delta_{pj} &= -\frac{\partial E_p}{\mathrm{net}_{pj}} \\
&= -\frac{\partial E_p}{\partial O_{pj}}\frac{\partial O_{pj}}{\partial \mathrm{net}_{pj}} \\
&= -(\sum_k \frac{\partial E_p}{\partial \mathrm{net}_{pk}}\frac{\partial}{\partial O_{pj}}\sum_i w_{ik}f(O_{pi}))k \\
&= k\sum_k \delta_{pk}w_{jk}f'(O_{pj})
\end{aligned}
\tag{5.12}
$$

since $\partial/\partial O_{pj}\sum_i w_{ik}f(Opi) = 0$ except when $i = j$.

Thus Equation 5.10 computes the gradient of the error function, and can be used to provide the first derivatives in optimisation methods which minimize the error function using algorithms such as gradient descent or conjugate gradient.

5.3.5 Kakadu MLP chip

Kakadu was fabricated using Orbit Semiconductor's $1.2\mu m$ double metal, single poly n-well process. A photomicrograph showing the main synapse blocks, row shift registers and the current source is shown in Figure 5.5. Kakadu has 10 input, 6 hidden and 4 output neurons and hence is called a 10:6:4 neural network. It can implement any smaller network than 10:6:4 by setting unused synapse values to zero. A summary of the major chip features is shown in Table 5.1.

Figure 5.5 *Photomicrograph of the Kakadu MLP chip.* © *1995 IEEE, reproduced with permission.*

5.4 Chip testing

5.4.1 Jiggle chip tester

The Kakadu chip was tested using the 'Jiggle' test jig [Leong and Vasimalla (1992)]. A photograph is shown in Figure 5.6. Jiggle was designed at the Systems Engineering and Design Automation Laboratory, Sydney University Electrical Engineering and is a general purpose chip tester having 64 12 bit analog input/output channels as well as 64 digital input/output channels. Jiggle connects to a VME bus, and the VME cage is interfaced to a Sun SPARC station IPC via a Bit 3 Model 466 SBUS to VME converter.

Jiggle allows arbitrary analog or digital signals to be presented to the pins of the test chip and thus allows software control of the weight updating and training of the Kakadu chip. For the experiments described below, a bias current of 6.63 nA and neuron values of 1.2 $M\Omega$ were used.

Figure 5.6 *Photograph of the Kakadu chip and Jiggle chip tester in VME cage.*

5.4.2 MDAC linearity test

The transconductance multiplier used in the Kakadu MDAC has a transfer function described by Equation 5.1. The output of the DAC (I_{DAC} in Figure 5.3) is equal to $I_{out+} - I_{out-}$ (in Equation 5.1) and so for fixed input voltages the equation can be simplified to

$$I_{out} = \alpha I_{DAC} \qquad (5.13)$$

where α is a constant and

$$I_{DAC} = \begin{cases} +\sum_{k=0}^{4} 2^k B_i & \text{if B5} = 1 \\ -\sum_{k=0}^{4} 2^k B_i & \text{if B5} = 0 \end{cases} \qquad (5.14)$$

I_{out} is connected to a neuron (a 1.2 MΩ pull-up resistor), and so the output can be measured as a voltage. Plots of the (measured) MDAC linearity for three bias currents are shown in Figure 5.7. One can see that monotonicity is not achieved for the bias current of 6.63 nA (at which the chip is operated). The two points at which it is not monotonic are when the absolute value of the DAC input

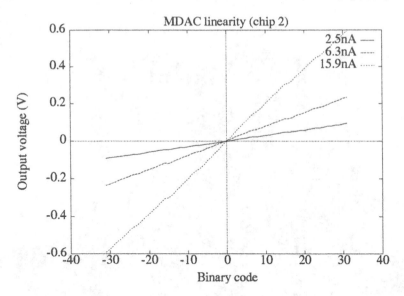

Figure 5.7 *MDAC linearity test (chip 2).*

changes from 15 to 16 and is hence due to the most significant bit current source. At 15.3 nA, however, the DAC is monotonic. Training can compensate for the nonlinearity of the DAC.

5.4.3 Synapse transfer function

The synapse followed by a neuron has a transfer function described by

$$V_{out} = R(I_{out+} - I_{out-}) \qquad (5.15)$$

where $R = 1.2 \times 10^6$ and $(I_{out+} - I_{out-})$ is given by Equation 5.1. A curve fit was used to find κ (26.0719) and a plot of the measured and expected synapse transfer function can be seen in Figure 5.8.

5.4.4 Power consumption

It is useful to be able to estimate the power consumption of a chip like Kakadu. This is a function which is linear with the weight values since I_{DAC} in Figure 5.3 is the current drawn for that particular synapse. A number of current consumption measurements were made for different weight values and then a least squares fit

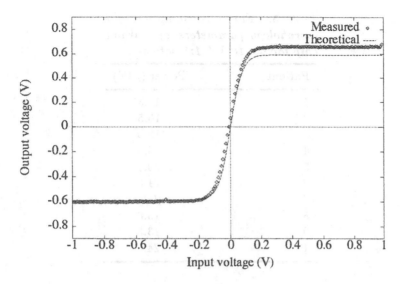

Figure 5.8 *Synapse transfer function.*

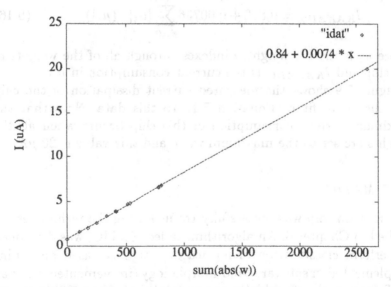

Figure 5.9 *Current consumption curve fit (Bias current = 6.63 nA).*

Table 5.2 *Power consumption and morphology parameters of Kakadu chip for the 10 VT 1:1 patients.*

Patient	Power (μW)
1	18.3
2	16.5
3	12.6
4	13.8
5	20.1
6	21.6
7	9.0
8	15.6
9	13.5
10	24.0

was used to derive the current consumption formula

$$I_{KAKADU} = 0.842 + 0.00736 \sum_{i=0}^{N} |w_i| \quad (\mu A) \qquad (5.16)$$

where w_i is the ith weight, i indexes through all of the weights in the chip and I_{KAKADU} is the current consumption in μA.

Figure 5.9 shows the measured current dissipation of the chip and the curve fit of Equation 5.16 to this data. Note that the maximum current consumption of this chip occurs when all the weights are set to the maximum value and this value is 20 μA.

5.4.5 MATIC

The Kakadu chip was successfully trained using the techniques described in Chapter 6. An algorithm, called MATIC, was developed to perform classification of intracardiac signals based on timing (implemented in software) and morphology (implemented using an artificial neural network) [Leong and Jabri (1992c,1993b)]. It was intended for use in an ICD.

The MATIC system was used on a database of 10 patients in order to identify arrhythmias which are only distinguishable by morphological features. A correct classification rate of 98.5% was achieved using the Kakadu chip. In order to compare the results of the experiment with that of a digital neural network, a standard two layer perceptron (TLP) of the same network size was first trained. The TLP network has marginally better performance (99.2%) than the Kakadu network and this is mostly due to the limited precision weight range available on the chip (6 bits).

The power consumption of Kakadu for the patients is shown in Table 5.2. The maximum power consumption of the chip was 25 μW for the patients studied. The propagation delay of the Kakadu chip is approximately 30 μs, a normal heart rate is approximately 1 Hz, and Kakadu has negligible ($<$ 100 pA) power consumption at zero bias. If a conservative value of 1000 μs for propagation is allowed, the average power consumption of the system can be reduced by a factor of 1000 to less than 25 nW by turning off the bias of the chip when it is not being used.

5.5 Discussion

Kakadu was designed to be used in implantable biomedical systems which require an analog interface and low power consumption. Compared with a digital approach, this system does not require analog to digital conversion, occupies small chip area and is very parallel. In contrast to other analog systems, implantable devices are not required to operate over a large temperature range.

The feasibility of using an analog VLSI connectionist's approach to address a power consumption constrained classification problem was demonstrated by successfully using the Kakadu chip to perform morphology classification on intracardiac electrograms.

CHAPTER 6

Supervised learning in an analog framework

6.1 Introduction

One of the main hurdles in the design and implementation of a microelectronic neural network is its training. Of course the difficulties in training an analog neural network depend on a number of factors, including:

- input/output representation: analog or binary
- size of network
- available weight resolution
- signal to noise ratio of the neurons, synapses and their associated signals.

From a training perspective, the ideal learning algorithm would provide robustness with respect to the factors above. During the last few years many analog neural network training algorithms have been reported. This chapter reviews several of these algorithms and discusses their advantages and drawbacks in the light of their practical performance in the training of an analog multi-layer perceptrons.

Although the experiments make use of multi-layer perceptrons only, the perturbation based algorithms we describe can be applied to other feed-forward architectures which can be trained using supervised learning. Some are also applicable to recurrent neural networks used for the recognition or estimation of sequential data.

6.2 Learning in an analog framework

The presentation in this chapter will make reference to an analog neural network 'chip', and in many cases, the network implementation will be referenced simply by the word 'chip'.

An analog multi-layer perceptron (AMLP) can be trained in three ways:

In-loop: The chip response is used in the weight optimisation process.

On-chip: Training is performed on chip.

Off-chip: Training is performed completely off the chip (*i.e.* no chip response is used) and the weights are downloaded to the chip (and truncated if required).

The process of in-loop training of AMLP chips is commonly used either as the sole chip training method or supplementary to off-chip training. The chip response includes its output as well as the activity of any hidden neurons, which may or may not be required by the training algorithm.* The training often requires the computation (usually off-chip) of either a Pattern Mean Squared Error (PMSE) or the error over all the patterns which we call here the Total Mean Squared Error (TMSE).

For the case of on-chip learning, there are a number of problems to be overcome. There have been several reports (see Hollis (1990); Hoelfeld and Fahlman (1991); Xie and Jabri (1992); Tarassenko, Tombs and Cairns (1993) for examples) indicating that techniques such as back-propagation require over 12 bits of resolution for the weights and states for successful training. Another drawback in the implementation of back-propagation on a chip is the complexity of the circuitry and the additional interconnections required for the backward propagation of the errors.

The difficulties in the implementation of back-propagation and its precision requirements have led to the investigation of algorithms that are simpler to implement and are less sensitive to precision. The algorithms on which most interest has focused so far have been based on the approximation of the error gradient by a finite difference. We will describe a collection of such finite difference based techniques in Section 6.5.

6.3 Notation

The notation used in the algorithm descriptions is as follows. The matrix $\mathbf{w}$ represents the network weights and its element w_{ij} represents the weight from neuron i to neuron j. The matrix $\Delta \mathbf{w}$

---

* It is assumed in this chapter that the weights of an MLP are stored on the host computer and down-loaded to the chip.

```
for all patterns {
    feed-forward;
    compute updates;
    apply updates;
}
```

Figure 6.1 *Normal on-line update strategy (NONLINE).*

```
for all patterns {
    feed-forward;
    compute updates;
}
apply updates;
```

Figure 6.2 *Normal batch update strategy (NBATCH).*

represents the update value of the weights. Unless otherwise indicated, the scalar $\varepsilon(\mathbf{w})$ represents the network error TMSE or PMSE. The scalar $p^{(n)}{}_{ij}$ represents the perturbation to be applied to the weight w_{ij} at iteration n of the learning process. The learning rate is represented by η.

6.4 Weight update strategies

In this chapter we make the distinction between learning algorithms and strategies. A learning algorithm refers to the weight update rule, while a learning strategy refers to the way the weight updates are applied to the weights (*e.g.* on-line, batch,...). In some cases, the algorithm and strategy are inherently linked. We will distinguish six different learning strategies:

1. normal on-line weight update (NONLINE): the weight updates are computed for each weight and for a given training pattern and are then applied. (see Figure 6.1).

2. normal batch mode weight update(NBATCH): the weight updates are computed for each weight and accumulated over all the training patterns and are then applied (see Figure 6.2).

```
for all patterns {
    feed-forward;
    measure PMSE;
    perturb weight(s);
    feed-forward;
    measure new PMSE;
    If (condition holds)
        keep perturbations;
}
```

Figure 6.3 *Single pattern conditional immediate update (CIONLINE).*

```
measure TMSE;
for all weights {
    perturb weight(s);
    measure new TMSE;
    If (condition hold)
        keep perturbations;
}
```

Figure 6.4 *Batch pattern conditional immediate update (CIBATCH).*

3. single pattern conditional immediate update (CIONLINE): Assuming that a weight update is available a weight is updated if a condition is met (e.g. error decreasing). If the condition evaluation involves error computation, then only the PMSE is evaluated (see Figure 6.3).

4. batch patterns conditional immediate update (CIBATCH): Similar to CIONLINE above with the exception that any error computation is done over all the patterns in the training set, that is the TMSE is computed (see Figure 6.4).

5. perturbation on-line weight update (PONLINE): the pattern mean squared error (PMSE) is measured; the weight(s) (depending on the perturbation parallelization scale) are perturbed and the new PMSE is measured; the weight updates are computed and applied. This is repeated for the next pattern and so on (see Figure 6.5).

```
for all patterns {
    measure PMSE;
    perturb weight(s);
    measure new PMSE;
    compute updates;
    remove perturbations;
    apply updates;
}
```

Figure 6.5 *Perturbation on-line update (PONLINE).*

```
feed-forward;
measure TMSE;
for all weight(s) {
    perturb weight(s);
    measure new TMSE;
    compute updates;
    remove perturbations;
}
apply updates;
```

Figure 6.6 *Perturbation batch update (PBATCH).*

6. perturbation batch (PBATCH): The total mean squared error (TMSE, over all patterns) is measured. The weights (depending on the parallelization scale) are perturbed and the new TMSE is measured. The weight updates are computed and applied (see Figure 6.6).

The weight update strategies described above can be applicable to a learning algorithm. The strategies PONLINE and PBATCH are targeted to sequential, semi-parallel and parallel weight perturbation. Although update strategies may seem trivial, in an in-loop or on-chip learning context, they have far reaching implications on the design and implementation aspects of an AMLP. Furthermore, batch strategies are impractical and do not scale well for perturbation based algorithms since they require the TMSE which is evaluated over all the training set. This requires additional storage and leads to a high computational cost per learning iteration.

6.5 Learning algorithms

In this section, learning algorithms that have been reported in the literature and used in the experiments are described. All algorithms reported in this chapter are part of the MUME simulation environment [Jabri, Tinker and Leerink (1994)].

6.5.1 Back-propagation

This is the standard back-propagation (BP) algorithm. The computation of the gradient (see Section 2.10 for the update rules) with respect to the weights is performed on the host computer, based on the readings of the chip's neuron outputs. The derivative of the neuron transfer function required by BP was based on a numerical model of the hardware developed by Leong and Jabri (1992b). It is expected that the numerical model of the derivative would introduce errors, but such errors would also be present and might be larger if an analog derivative circuit implementation was used. The update strategies used for the BP algorithm are NONLINE and NBATCH.

6.5.2 Weight perturbation

The gradient with respect to the weight can be evaluated using the gradient approximation method of Forward Finite Difference [Jabri and Flower (1992)] (see Figure 6.7). We call this method Weight Perturbation (WP). The weight updates are computed as

$$\Delta w_{ij} = -\eta \frac{\varepsilon \left(w + p^{(n)}\right) - \varepsilon(w)}{p_{ij}^{(n)}} \qquad (6.1)$$

Note, here only the element p_{ij} of the matrix $\mathbf{p^{(n)}}$ is non-null. The error ε represents either the TMSE or the PMSE. As only one weight is perturbed at any one time, the algorithm is sequential with respect to the weights. This is the simplest of the weight perturbation techniques which aims to compute a finite difference approximation to the gradient of the error with respect to the weight.

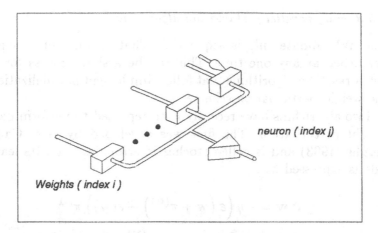

Figure 6.7 *Weight Perturbation. A perturbation is injected at the weight and the difference of the network's error before and after the injection is used to approximate the error's gradient.*

The update strategies used in the experiments are PONLINE and PBATCH.

The order of the error of the forward finite difference approximation can be improved by using the Central Difference method. However, the number of forward relaxations of the network required for the Central Difference method is twice that required for the Forward Difference which we have shown above.

WP is then equivalent to any other gradient descent method with the variation that the search direction is generated on a per weight basis using a Finite Difference method, and should not be confused with a 'Blind Search' method. As with all gradient descent methods, WP is susceptible to being trapped in local minima, however, the various techniques used for escaping or avoiding these in other gradient descent optimisation techniques are also applicable.

WP should not be confused with the 'model free' learning approach proposed by Dembo and Kailath (1990) which make use of continuous perturbation of the weights while a time varying performance index is measured and correlated with the perturbation signals. The key feature of this method is the approximation of the instantaneous gradient of the actual performance index, which is accomplished by parallel weight perturbations followed by local correlations.

6.5.3 Fully parallel perturbation algorithms

The WP update rule is sequential. That is, only one weight is perturbed at any one time. This can be a slow process for large networks. The algorithms that follow implement parallelizations of the weight perturbation scheme.

Two algorithms have recently been reported to perform parallel weight perturbation. The first was developed by Gert Cauwenberghs (1993) and is called stochastic error descent . Its learning rule is expressed as

$$\Delta \mathbf{w} = -\eta \left(\varepsilon \left(\mathbf{w} + \pi^{(\mathbf{n})} \right) - \varepsilon(\mathbf{w}) \right) \pi^{(\mathbf{n})} \qquad (6.2)$$

where an element $\pi_{ij}^{(n)}$ of the matrix $\pi^{(\mathbf{n})}$ corresponds to a perturbation $p_{ij}^{(n)}$. Note that all weights in the net are perturbed simultaneously and that the elements $\pi_{ij}^{(n)}$ of the perturbation matrix $\pi^{(\mathbf{n})}$ have to be orthogonal (spatially and temporally uncorrelated)

$$E \left(\pi_{ij}^{(n)} \pi_{kl}^{(m)} \right) = \sigma^2 \delta_{(ij,kl)} \delta_{(n,m)} \qquad (6.3)$$

where E denotes expectation, σ is the perturbation intensity and δ represents the Kronecker Delta symbol

$$\delta_{(ij,kl)} = \begin{cases} 1 & \text{if } i = k \text{ and } j = l \\ 0 & \text{otherwise} \end{cases} \qquad (6.4)$$

and the Kronecker Delta symbol $\delta_{(n,m)}$ is defined as

$$\delta_{(n,m)} = \begin{cases} 1 & \text{if } n = m \\ 0 & \text{otherwise} \end{cases} \qquad (6.5)$$

If the condition of Equation 6.3 is not met, then greater credit assignment confusions will occur which will lead to a deterioration in the training efficiency. We will elaborate more on this issue in Section 6.6.

In an analog limited precision environment, the perturbation strength will have a minimum which is equal to the smallest change that can be applied to a weight distinguishable from noise. In practice, the algorithm can be reduced to the case where the perturbation magnitude is a small constant and the perturbation sign is randomly generated. This leads to the second fully parallel perturbation algorithm developed independently by Josh Alspector and colleagues [Alspector, Meir, Yuhas, Jayakumar and Lippe (1993)] which we call here constant perturbation random sign (CPRS). Its learning rule can be expressed as

$$\Delta \mathbf{w} = -\eta \left(\varepsilon \left(\mathbf{w} + \mathbf{p^{(n)}} \right) - \varepsilon(\mathbf{w}) \right) \mathbf{p'^{(n)}} \qquad (6.6)$$

where $p^{(n)}$ is the matrix of the perturbations p_{ij} and $p'^{(n)}$ is a matrix of the perturbation elements $1/p_{ij}$ (inverse of the perturbations). All the perturbations $p_{ij}^{(n)}$ are equal in strength but random in sign. All weights in the net are perturbed simultaneously. The update strategies used are PONLINE and PBATCH.

6.5.4 Summed weight neuron perturbation

The Summed Weight Neuron Perturbation (SWNP) algorithm is similar in form to CPRS and was developed independently by Flower and Jabri (1993a). It is a compromise between sequential and parallel weight perturbation in that only the weights feeding into one neuron are perturbed in parallel (see Figure 6.8). This means that several sequences of perturbations are required for the complete net.

The SWNP learning rule iterates over all the neurons. For neuron j the expression of the learning rule is

$$\Delta \mathbf{w}_{\rho_{\mathbf{j}}} = -\eta \left(\varepsilon \left(\mathbf{w} + \rho_{\mathbf{j}}^{(\mathbf{n})} \right) - \varepsilon(\mathbf{w}) \right) \rho_{\mathbf{j}}'^{(\mathbf{n})} \qquad (6.7)$$

where $\rho_{\mathbf{j}}^{(\mathbf{n})}$ is a matrix where only the elements corresponding to the connections feeding into neuron j are non-null. The element $\rho_{ij}^{(n)}$ in this matrix corresponds to a perturbation $p_{ij}^{(n)}$. The matrix $\rho_j'^{(n)}$ has the inverse of the perturbations $(1/p_{ij}^{(n)})$. Note that in form the rule is similar to CPRS with the main difference being in the *scheduling* of the perturbations. The update strategies used are PONLINE and PBATCH.

6.5.5 Combined local / random search algorithms

The perturbation algorithms we have presented above (sequential or parallel) are local in their search, that is, they follow the gradient (or its approximation) of an error with respect to the parameters being optimized. When the weights are quantized, the implications are that the error surface will present discontinuities and areas of zero gradient. These will affect the learning algorithms which

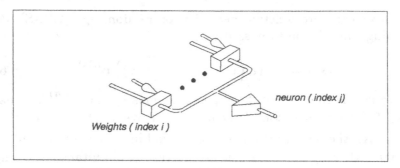

Figure 6.8 *Summed weight neuron perturbation (also called Fan-In Perturbation). Perturbations of random values are injected simultaneously at all weights feeding into a neuron (j here).*

may become unstable and/or stuck in local minima [Jabri, Pickard, Leong and Xie (1993); Xie and Jabri (1992a)].

The instability results from the failure of the learning algorithm to evaluate small enough moves on the error surface because of the quantisation resolution. If the resolution cannot be improved, the only available choice is to search the solution space for another state which has non-zero gradient for the given weight resolution. This is difficult to achieve using local search techniques. The algorithms we review below address these problems.

Combined Search Algorithm (CSA)

This algorithm is based on [Xie and Jabri (1992)]. It performs a combination of modified weight perturbation and a partial random search. Modified Weight Perturbation (MWP) is a local learning rule that tests for a drop in the TMSE before adopting an immediate weight change. Partial random search is a global rule that randomly changes a weight and adopts the change if a reduction of the TMSE is observed. The update strategy used is CIBATCH.

Simulated Annealing (SA)

A temperature parameter T is used to anneal the learning process. The temperature and the TMSE are used to produce a move acceptance threshold function $MAT(T, TMSE)$. The simulated annealing weight update rule randomly changes a weight and the change is adopted either if the TMSE is reduced or if a randomly generated number R satisfies the relation

$$R \geq MAT\left(T, TMSE_{w_{new}} - TMSE_{w_{old}}\right)$$

The update strategy used is CIBATCH.

Combined local search and simulated annealing (MWP-SA)

This method alternates the use of a modified weight perturbation with simulated annealing. The update strategy used is CIBATCH.

It is important to note that CSA, SA and MWP-SA do not scale well, meaning that they are inefficient if the network being trained is large.

6.6 Credit assignment efficiency

There are two main motivations behind the parallelization of weight perturbation. The first, and the obvious, is to explore the weight solution space faster. The second motivation is to introduce some diffusion in the learning process which could assist in the avoidance of local minima. However, the outcome must be seen in light of the efficiency of the credit assignment, that is, whether the training time (number of feed-forwards) is effectively reduced and whether the quality of the training solution is maintained (consistency in training convergence and generalization performance). We argue that in practice, fully parallel perturbation algorithms can be much less efficient than semi-parallel algorithms. We do not prove this theoretically, but we will provide an intuitive analysis and later present extensive experimental results to support this hypothesis.

A proof that fully parallel weight perturbation algorithms can perform gradient descent on average can be found in [Cauwenberghs (1993)]. The reduction in computational complexity for each weight update of fully parallel perturbation methods with respect to sequential weight perturbation is a factor of $\sqrt{N}$ where N is the number of parameters [Cauwenberghs (1993)]. A description of SWNP demonstrating the $O(\sqrt{N})$ reduction in complexity over sequential weight perturbation was presented in [Flower and Jabri (1993a)] together with the proof that, for SWNP, the descent steps in the direction of the negative of the gradient were larger in size (regardless of the learning rate) than the steps in the wrong direction, and hence that SWNP does implement an error descent scheme.

Semi-parallel and fully parallel weight perturbation algorithms often introduce credit assignment confusions. A credit assignment confusion is the situation where the update to a weight (or collection of weights) leads to an increase in the error regardless of the size of the learning rate. A simple illustrative example of credit assignment confusion is the situation where the output state of a neuron in a network is zero and yet the weights being driven by that neuron are changed by the updates. This can often happen in semi- and fully parallel weight perturbation schemes. However, in algorithms that compute the gradient analytically (using the chain rule like back-propagation) these weights will not change if the state of the feeding neuron is zero. Confusions can also exist if the condition of Equation 6.3 is not met. Although such confusions can take place in semi-parallel and parallel perturbation schemes, we argue that the reasons why SWNP in general leads to more efficient learning in the case of AMLPs are:

1. In fully parallel perturbation schemes such as those of [Cauwenberghs (1993)] and [Alspector, Meir, Yuhas, Jayakumar and Lippe (1993)], credit assignment confusions are introduced regardless of whether the perturbing signals are totally uncorrelated or not. This is a direct implication of the proof in [Cauwenberghs (1993)]. Some perturbations will lead to bad credit assignment, and some will lead to good ones. There is no argument that the number of good assignments is on average higher and hence gradient descent takes place. The question is how efficient the credit assignment process is. In SWNP all the weights being perturbed feed into a single neuron. Confusions in credit assignment as expected in fully parallel perturbations are contained or localised within the fan-in neuron.

2. Due to limited precision, noise and perturbation injection implementations in AMLPs, the smallest perturbation is equal to the smallest quantity that can be injected into the weights and be distinguished from correlated noise. Smallest perturbations would favour fully parallel schemes [Cauwenberghs (1993)]. The design of circuitry that will allow high precision perturbation and computation in analog is itself a major engineering project. The experiments presented here indicate that semi-parallel perturbation schemes tolerate larger perturbations than fully parallel weight perturbation schemes.

We are not claiming that partially parallel perturbation techniques such as SWNP will always out-perform fully parallel perturbation techniques, but that conditions exist for semi-parallel schemes to be more effective.

6.7 Parallelization heuristics

In this section we describe two new parallelization heuristics: FAN-OUT and FAN-IN-OUT. The proof that both these techniques do perform gradient descent on average is a trivial case of the proof provided for the SED algorithm in [Cauwenberghs (1993)].

6.7.1 FAN-OUT semi-parallel perturbation

The technique we describe in this section can be called a fan-out technique (in contrast with the fan-in SWNP) and is depicted in Figure 6.9 In a fan-out technique all the weights leaving a neuron are perturbed simultaneously using uncorrelated random values. The learning rule is

$$\Delta w_{\xi_j} = -\eta \left(\varepsilon \left(w + \xi_j^{(n)} \right) - e(w) \right) \xi'_j^{(n)} \qquad (6.8)$$

where $\xi_j^{(n)}$ is a matrix where only the elements corresponding to the connections leaving a neuron j are non-null. The element $\xi_{ij}^{(n)}$ in this matrix corresponds to a perturbation $p_{ij}^{(n)}$. The matrix $\xi'_j^{(n)}$ has the same structure as $\xi_j^{(n)}$ except that its elements correspond to the inverse of the perturbations $(1/p_{ij}^{(n)})$. The advantage of fan-out over a fan-in technique (i.e., SWNP) can be described as follows. Since all the weights leaving a neuron will feed either to neurons in the following layer or to the outputs, a wider range of the solution space is explored. In a fan-in technique, the range can be constrained by the fact that only one neuron s state is effectively being affected. This at first glance may appear in conflict with the containment argument described in Section 6.6. It is true that the containment effect is lost by a fan-out method, but an important gain (probably more advantageous than the containment) is achieved and that is, all the weights being perturbed are being fed with the same neuron state. As a result, credit assignment confusions due to varying neuron state as in SWNP and fully parallel perturbations are removed. The experiments later show a

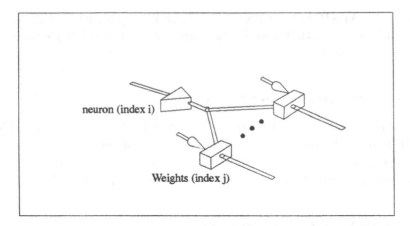

Figure 6.9 *FAN-OUT based partial perturbation.*

significant improvement in training performance for the fan-out method over all other methods described so far in this chapter.

6.7.2 *FAN-IN-OUT semi-parallel perturbation*

In this semi-parallel perturbation scheme, all the weights feeding and leaving a neuron are simultaneously perturbed using uncorrelated random values (see Figure 6.10). The learning rule is

$$\Delta \mathbf{w}_{\psi_{\mathbf{j}}} = -\eta \left(\varepsilon \left(\mathbf{w} + \psi_{\mathbf{j}}^{(\mathbf{n})} \right) - \mathbf{e}(\mathbf{w}) \right) \psi_{\mathbf{j}}'^{(\mathbf{n})} \qquad (6.9)$$

where $\psi_j^{(n)}$ is a matrix where only the elements corresponding to the connections feeding (index i in Figure 6.10) and leaving (index k in Figure 6.10) neuron j are non-null. An element $\psi_{ij}^{(n)}$ in this matrix corresponds to a perturbation $p_{ij}^{(n)}$. The matrix $\psi_j'^{(n)}$ has the same structure as $\psi_j^{(n)}$ but its elements correspond to the inverse of the perturbations $(1/p_{ij}^{(n)})$. Clearly, the FAN-IN-OUT heuristic is a combination of SWNP and FAN-OUT. Although one cannot expect that the full advantages of SWNP and FAN-OUT will be achieved in FAN-IN-OUT, we expect that learning efficiency would be as good as FAN-OUT and the reduced computational complexity of exploring the fan-in weights will still be maintained.

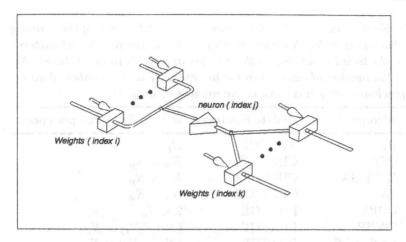

Figure 6.10 *FAN-IN-OUT based partial perturbation.*

6.7.3 Epoch time cost

An epoch is an iteration through a training set. We show in Table 6.1 the relationship between the number of epochs and the number of feed-forwards for each of the learning algorithms, assuming a batch type update strategy is used. In the case of on-line update strategies, the relationship is shown in Table 6.2 for the relevant algorithms. In both tables we have neglected the differences in the amount of computation required to modify the weights for the perturbation based algorithms, which according to the experience does not significantly affect the time cost of an epoch.

6.8 Experimental methods

Experiments on two problems were conducted: intracardiac electrogram classification and the 4 dimensional parity learning problem. The selection of these two problems was a matter of data convenience and the architecture (size) of the network implemented on the chip.

Each of the experiments was repeated at least 10 times, and some were repeated 20 times. Clearly the rate at which experiments can be performed is limited by the fact that only one hardware set up was available (described below) and the training bottleneck which is the bus interface between the chip and the computer (as the training is being performed in-loop). As a result, not all combi-

Table 6.1 *Number of feed-forwards per epoch for each of the training algorithms in batch update strategies. N_p is the number of patterns in the training set. N_w is the number of weights in the network. N_n is the number of neurons in the network. N_i is the number of inputs (including any bias). N_o is the number of outputs.*

Algorithm	Update Strategy	# Feed-Forwards per epoch
BP	NBATCH	N_p
CSA	CIBATCH	$N_w \times N_p$
MWP-SA	CIBATCH	$N_w \times N_p$
SA	CIBATCH	$N_w \times N_p$
CPRS	PBATCH	$2 \times N_p$
SWNP	PBATCH	$(Nn - N_i) \times N_p$
FAN-OUT	PBATCH	$(Nn - N_o) \times N_p$
FAN-IN-OUT	PBATCH	$(Nn - N_i - N_o) \times N_p$
WP	PBATCH	$N_w \times N_p$

Table 6.2 *Number of feed-forwards per training pattern for each of the training algorithms in on-line update strategies. The descriptions of the parameters are shown in the caption of Table 4.1*

Algorithm	Update Strategy	# Feed-Forwards per pattern
BP	NONLINE	1
CPRS	PONLINE	2
SWNP	PONLINE	$N_n - N_i + 1$
FAN-OUT	PONLINE	$N_n - N_o + 1$
FAN-IN-OUT	PONLINE	$N_n - N_i - N_o + 1$
WP	PONLINE	$N_w + 1$

nations of training strategies/algorithms/task were performed. We show the experiments that were performed in Table 6.3.

Table 6.3 *Summary of the experiments. The weight and weight update on the computer column has two cases: floating point (FP) and both weight updates and weights quantised (Q). Note that all the weights are quantised before being loaded onto the Kakadu chip(see Section 6.8.1) regardless of the way they have been modified on the computer.*

Training Algorithm	Update Strategy	ICEG		Parity 4	
		FP	Q	FP	Q
BP	NBATCH	✓			
	NOLINE	✓			
CSA	IBATCH	✓			
MWP-SA	IBATCH	✓			
SA	PBATCH	✓			
CPRS	PONLINE	✓	✓	✓	✓
	PBATCH	✓			
SWNP	PONLINE	✓	✓	✓	✓
	PBATCH	✓			
FAN-OUT	PBATCH	✓			
	PONLINE	✓		✓	✓
FAN-IN-OUT	PONLINE	✓	✓	✓	✓
WP	PBATCH	✓			
	PONLINE	✓	✓	✓	✓

6.8.1 The analog MLP

The AMLP chip used in the experiments is called KAKADU and was described in Chapter 5. This chip has a 10-6-4 architecture[†] and is built using the MDAC synapses described in Section 4.2.1 in Chapter 4. The weights are stored on the chip as 6-bit numbers (5 plus the sign) in a matrix of static registers. The chip is interfaced

---

† Depending on the experiments, unused weights for a particular architecture are set to 0.

to a host computer by means of an analog/digital board (called Jiggle) located in a VME cage.

In all the experiments the weights and weight updates were maintained on the host and were downloaded to the chip to perform the feed-forward.

6.8.2 Intracardiac electrogram data set

These data are classifications of a patient's intracardiac electrogram (ICEG) signals. The input component of each training vector is 10 samples of the right ventricular apex (RVA) lead. The output component is an indication of whether the heart rhythm is a normal sinus rhythm (NSR) or a retrograde ventricular tachycardia (VTR). The data was recorded at an electro-physiological studies (EPS) laboratory and has been normalised using a fixed scale to allow a range between -200mV and 200mV. For each experiment 8 training patterns and 220 testing patterns were used. The 220 test patterns are divided into 150 VTR and 70 NSR. Note the small number of training patterns is not surprising as the network's computation is limited in precision. Larger training sets do not improve the generalization performance and only lead to longer training time. The network architectures used in the experiments are 10-5-1 for the batch strategies and 10-6-1 for the on-line strategies.

6.8.3 Parity 4 data set

For Parity 4, the task is to classify the binary 4-dimensional input vectors (16 of them) according to their parity (even or odd). A typical MLP that solves this problem (in floating point simulation) consists of a 5-4-1 with one of the inputs used as a bias. In the experiments we report below, an architecture of 5-5-1 or 5-6-1 was used. A binary '1' corresponds to an input to the chip of 200mV and a binary '0' corresponds to 0V. The 5-5-1 architecture was used when the weights were maintained (on the host computer) in floating point. The 5-6-1 architecture was used when the weights on the host computer were maintained in truncated form (Section 6.9.2 describes the truncation procedure).

6.8.4 Training convergence

Each training session is terminated if either the maximum number of iterations (MAXITER) is reached or the TMSE has reached a critical error value (CEV) for which the learning process would be considered to have 'converged'. In many of the experiments, learning algorithms would reach TMSE values very close to CEV but fail to 'converge' within the maximum number of iterations. When training reaches 100% of patterns correct, the training is considered to have converged, regardless of whether it has reached CEV or MAXITER.

6.8.5 Learning rate

In most experiments, the learning rate had to be optimised for the algorithm. The single output of the KAKADU chip can present a maximal mean square error of about 0.002. Obviously if one has truncated the weight updates (to make them integers between -31 and +31) a large learning rate is required. This however may lead to the training 'oscillating'. Trial and error was used to determine appropriate rates for the learning algorithms. The rate that was found to be the most successful was then selected. In some experiments (especially those with truncated weights and weight updates) up to one hundred pre-trials have been performed to determine appropriate learning rates. Overall, the experiments took over 14 months to complete!

6.8.6 Perturbation strength

In all the experiments reported in this chapter, the strength of the perturbation was set to 1. The strength of the perturbation is an important parameter for perturbation algorithms. Like digital MLPs, weight storage in AMLPs will be subject to quantisation regardless of the storage technology. Of course the effects of quantisation could be reduced (but not indefinitely) by using more sophisticated storage circuitry at the cost of an increased chip area. The perturbations applied to the weights of an AMLP are larger than or equal to the minimum quantum that the circuitry can manage. In the case of KAKADU where the weights are stored in digital and converted to analog, the minimum perturbation magnitude should be one. This is not a limitation of KAKADU style weight storage. Other implementation methods such as those described in [Cauwenberghs (1994)] and [Flower, Jabri and Pickard (1994)] have equivalent constraints on the minimum perturbation strength.

During in-loop training, the perturbation strength could be reduced if the weights are maintained in floating point format (and truncated when down-loaded to the chip). This is due to the accumulation of small weight perturbations made possible by a floating point representation. From the experiments, however, we do not find any improvements, but on the contrary, the training sessions required a much larger number of epochs.

6.9 ICEG experimental results

In this section the results for training and generalization on the ICEG data base are presented. Both batch and online modes are considered.

6.9.1 Batch mode

The batch mode set of experiments were performed on the ICEG classification problem only. A 10-5-1 AMLP architecture was used. The weights and weight updates were maintained in floating point representations. The weights are quantized to 6 bits before being loaded into the chip to produce its response to an input pattern. For each algorithm 20 training trials were performed. Each trial starts with different initial conditions. An upper limit of 2000 training iterations was set. This upper limit was determined by:

- excessive experimental time required by some algorithms,
- average performance of most algorithms

We show the training performance of the eight algorithms in Table 6.4. The first column shows the algorithm; the second column shows the update mode; the third column has two sub-columns showing the average final training error and its standard deviation respectively (on the runs that have successfully trained); the last column shows the number of times an algorithm has successfully trained all the patterns (i.e. converged) out of the 20 runs. We show in Table 6.5 the computational costs of the eight algorithms. The second column shows the average and standard deviation of the number of epochs for the runs that have successfully trained.

Table 6.4 *Training performance summary of the learning algorithms on the ICEG classification problem for 20 trials. See text for explanation.*

Algorithm	Update mode	Training TMSE		Converged
Max Iter:2000 CEV:0.0004		Average	Standard deviation	
BP	NBATCH	3.48e-04	5.96e-05	14
CPRS	PBATCH	3.46e-04	5.50e-05	12
CSA	CIBATCH	3.36e-04	4.80e-05	20
MWP-SA	CIBATCH	3.50e-04	5.30e-05	20
SA	CIBATCH	3.16e-04	8.09e-05	20
SWNP	PBATCH	3.66e-04	3.65e-05	18
FAN-OUT	PBATCH	3.82e-04	2.06e-05	20
WP	PBATCH	3.57e-04	2.61e-05	20

The third column also has two sub-columns showing the average and standard deviation of the number of feed-forwards performed during training. Note these numbers are related to the number of epochs using the information of Table 6.1.

Table 6.4 shows that all algorithms have a probability of more than 50% of converging. Table 6.5 shows that epoch wise, CSA is the fastest. Based on the number of feed-forwards, BP is the fastest followed by CSA. While BP has fewer feed-forwards, it is more computationally expensive than the other algorithms when computing the weight updates. Note that all algorithms except BP, CPRS and SWNP, managed to converge in all 20 trials.

The generalization (or testing) results are shown in Table 6.6. The first column in the table shows the algorithm; the second column has two sub-columns showing the average and the standard deviation of the TMSE over the runs that have successfully trained. The smaller the standard deviation the more consistent the performance of the network was across the successful runs. The third column shows the classification performance (for the successful runs), again with a column showing the average and another column showing the standard deviation. Most algorithms generalize

Table 6.5 *Training cost summary of the learning algorithms on the ICEG classification problem for 20 trials. See text for explanation.*

Algorithm	Epochs		Feed-forwards	
Max Iter:2000 CEV:0.0004	Average	Standard deviation	Average	Standard deviation
BP	960	329	7680	2632
CPRS	1391	216	22256	3456
CSA	35	18	15400	7920
MWP-SA	88	81	38720	35640
SA	165	85	72600	37400
SWNP	1270	277	60960	13296
FAN-OUT	1165	323	139800	38760
WP	466	162	205040	71280

pretty well, with a cluster of them performing in the high nineties. The best average performance was achieved by CPRS followed very closely by FAN-OUT and SWNP. The number of successful runs of SWNP and CPRS is however smaller.

6.9.2 On-line mode

In the case of on-line training, two sets of experiments were performed:

1. weights and weight updates in floating point.

2. weights and weight updates truncated to 6 bits.

For the ICEG on-line mode experiments a 10-6-1 AMLP architecture was used. The 10-5-1 architecture used in the batch mode experiments was evaluated and found to be inferior to the 10-6-1 architecture (within the constraint of the maximum number of iterations). The algorithms CSA, MWP-SA and SA were not experimented with as earlier work has shown very poor performance when on-line update modes are used.

Table 6.6 *Generalization performance summary of the learning algorithms on the ICEG classification problem (batch mode).*

Algorithm	Testing error(TMSE)		Classification (% correct)	
	Average	Standard deviation	Average	Standard deviation
BP	4.31e-04	2.56e-04	95.62	4.56
CPRS	2.01e-04	8.8e-05	96.97	2.29
CSA	3.27e-04	3.92e-04	95.57	5.55
MWP-SA	1.08e-03	1.06e-03	86.89	12.98
SA	7.70e-04	5.81e-04	90.72	7.45
SWNP	2.22e-04	1.20e-04	96.21	3.27
FAN-OUT	2.06e-04	1.3e-04	96.86	2.93
WP	3.80e-04	1.69e-04	94.48	2.99

Weights and weight updates in floating point

In this set of experiments, 20 trials were conducted for each algorithm. The weights were maintained on the host in floating point representation (double precision) and were truncated to integers in the range [-31, 31] when used on the chip. A summary of the training performance of the algorithms is shown in Table 6.7 and the computational cost in Table 6.8. Note the poor performance (in terms of number of successful runs) of BP. All perturbation based algorithms have a respectable convergence rate. The algorithm with the highest probability of successful training is FAN-IN-OUT, only failing 2 out of the 20 runs. Interestingly, FAN-IN-OUT also has a competitive feed-forward average and standard deviation when compared to BP and CPRS.

The generalization performance of the algorithms is shown in Table 6.9. The best average generalization performance is achieved by FAN-IN-OUT with an average correct classification of 98.61% and a very small standard deviation. FAN-OUT achieves a very close average performance and a very small standard deviation as well. All the algorithms in fact achieve an average performance

Table 6.7 *Training performance on the ICEG classification problem in on-line update mode. Weights and weight updates are maintained in floating point representation.*

Algorithm	Update mode	Training TMSE		Converged
Max iter:1000 CEV:0.0008		Average	Standard deviation	
BP	NONLINE	6.65e-05	1.56e-05	4
CPRS	PONLINE	6.62e-05	1.50e-05	11
SWNP	PONLINE	7.04e-05	1.23e-05	11
FAN-OUT	PONLINE	7.57e-05	3.65e-06	16
FAN-IN-OUT	PONLINE	6.98e-05	5.87e-06	18
WP	PONLINE	7.08e-05	6.23e-06	12

Table 6.8 *Training cost on the ICEG classification problem in on-line update mode. Weights and weight updates are maintained in floating point representation.*

Algorithm	Epochs		Feed-forwards	
Max iter:1000 CEV:0.0008	Average	Standard deviation	Average	Standard deviation
BP	919	63	919	63
CPRS	810	227	1620	454
SWNP	597	242	4776	1936
FAN-OUT	662	100	11254	1700
FAN-IN-OUT	701	209	4907	1463
WP	633	224	42411	15008

Table 6.9 *Generalization performance of the algorithms on the ICEG classification problem (on-line mode).*

Algorithm	Testing error		Classification (% correct)	
	Average	Standard deviation	Average	Standard deviation
BP	3.45e-04	1.13e-04	97.39	0.78
CPRS	2.17e-04	1.01e-04	97.48	1.96
SWNP	2.84e-04	1.80e-04	96.49	3.46
FAN-OUT	1.98e-04	6.71e-05	98.18	1.17
FAN-IN-OUT	1.67e-04	7.26e-05	98.61	1.46
WP	3.18e-04	1.03e-04	97.76	1.28

in the high 90s. The generalization performance of BP has to be treated carefully as it is based on the 4 successful trials only.

Truncated weights and weight updates

In this set of experiments, both the weights and their updates are maintained in a 6-bit integer format [-31 to +31]. The truncation process is performed as in Figure 6.11. Note the weight updates are first multiplied by the learning rate; they are then quantised to six bits; they are added to the weights; and then the weights are quantised to six bits. This particular way of quantising the weights and their updates is 'implementation' motivated. In an AMLP, the weight updates are most likely to be implemented as charge or discharge of capacitors (see [Alspector, Meir, Yuhas, Jayakumar and Lippe (1993); Cauwenberghs (1994) and Leong (1992a)] for examples) and hence will be applied as a 'quantum'. The weights themselves have to be either maintained in a digital random access memory (RAM), electrically erasable programmable read only memory (EEPROM) or refreshed from their analog values. In any of these cases, weights will be truncated to a limited precision.

Unlike the previous experiments, only 10 trials were conducted and the the maximum number of iterations has been increased to 2000 and convergence criteria was relaxed to 0.0001. Due to

```
foreach weight[i,j]{
    w[i,j] = w[i,j] + Quantise(Eta * dw[i,j]);
    w[i,j] = Quantise(w[i,j]);
}
```

Figure 6.11 *Truncation of weight updates and weights.*

Table 6.10 *Summary of the 10 training trials on the ICEG data with truncated weights and weight updates. The update strategy PONLINE was used.*

Algorithm	Training error		Converged
Max iter: 2000 CEV: 0.0001	Average	Standard deviation	
CPRS	8.16e-05	1.83e-05	5
SWNP	8.53e-05	8.8e-06	7
FAN-OUT	9.48e-05	6.29e-06	10
FAN-IN-OUT	9.33e-05	5.86e-06	10
WP*	8.70e-05	1.31e-05	10

* As in the previous experiment, WP's max number of iterations was set to 1000.

the poor performance of BP in the previous experiment, it was excluded from this experiment. The training results are shown in Table 6.10 and Table 6.11. Note that only FAN-OUT, FAN-IN-OUT, and WP managed to converge 10 times out of the 10 trials. This is followed by SWNP which manages to converge 7 times and CPRS which converges 5 times. Interestingly, FAN-IN-OUT is not only the most consistent, but is also the fastest epoch and feed-forward wise.

The generalization performance of the algorithms is shown in Table 6.12. Note if one neglects that the sample size (number of successful runs) of CPRS is half that of FAN-OUT, FAN-IN-OUT and WP, CPRS and SWNP achieve the best average generalization performance with SWNP being the most consistent (over 7 runs

Table 6.11 *Computational cost of the 10 training trials on the ICEG data with truncated weights and weight updates.*

Algorithm	Epochs		Feed-forwards	
Max iter: 2000 CEV: 0.0001	Average	Standard deviation	Average	Standard deviation
CPRS	1517	357	3034	714
SWNP	872	517	6976	4136
FAN-OUT	667	622	11339	10574
FAN-IN-OUT	409	434	2863	3038
WP	920	616	61640	41272

Table 6.12 *Generalization performance on the ICEG classification problem with truncated weights and weight updates.*

Algorithm	Testing error		Classification (% correct)	
	Average	Standard deviation	Average	Standard deviation
CPRS	1.48e-04	6.09e-05	98.64	1.79
SWNP	1.94e-04	1.36e-04	98.31	1.28
FAN-OUT	3.4e-04	2.13e-04	95.77	4.27
FAN-IN-OUT	3.43e-04	2.99e-04	95.41	5.28
WP	6.59e-04	7.53e-04	92.09	8.29

of course). In the group of algorithms that converged 10/10, FAN-OUT achieves the best and most consistent performance, followed very closely by FAN-IN-OUT.

6.10 Parity 4 experimental results

For the parity problem only on-line training was performed. As for the ICEG problem two weight update methods were used:

1. weights and weight updates in floating point.

Table 6.13 *Summary of the 20 training trials on the Parity 4 problem. The update strategy PONLINE was used.*

Algorithm	Training error		Converged
Max iter: 2000 CEV: 0.00008	Average	Standard deviation	
CPRS	4.37e-04	4e-04	5
SWNP	7.50e-05	4.33e-06	12
FAN-OUT	7.26e-05	6.29e-06	14
FAN-IN-OUT	1.02e-04	9.16e-05	10
WP	7.6e-05	2.83e-06	6

2. weights and weight updates truncated to 6 bits.

6.10.1 Weights and weight updates in floating point

In this experiment a 5-5-1 architecture was used. A 5-4-1 architecture was tried but could not lead to 100% correct training. Table 6.13 shows the training performance summary and Table 6.14 shows the computational cost for 20 training trials. The best performance is achieved by FAN-OUT followed by SWNP. FAN-OUT is also the most consistent. Note that FAN-IN-OUT has achieved a respectable performance when compared to WP and CPRS.

As the Parity 4 problem is more of a training challenge (rather than a generalization challenge as it is the case with the ICEG classification problem), we show in Table 6.12 the average performance of the training algorithms over all the trials (not only the successful ones as we did in the ICEG experiments). The experiments also show that FAN-OUT performs best on average with the highest correct classification rate and smallest standard deviation over all 20 trials. The results also show a consistency between the performance (average) of the successful trials and the performance over all the trials.

Table 6.14 *Computational cost of the 20 training trials on the Parity 4 problem. The update strategy PONLINE was used.*

Algorithm	Epochs		Feed-forwards	
Max iter: 2000 CEV: 0.00008	Average	Standard deviation	Average	Standard deviation
CPRS	1995	11	3990	22
SWNP	970	406	7760	3248
FAN-OUT	1069	273	12828	3276
FAN-IN-OUT	1453	414	10171	2898
WP	1083	482	40071	17834

Table 6.15 *Overall training performance on all trials for the Parity 4 problem (including trials that did not converge).*

Algorithm	Testing error		Classification (% correct)	
	Average	Standard deviation	Average	Standard deviation
CPRS	6.61e-04	4.24e-04	91.88	6.76
SWNP	3.14e-04	4.01e-04	96.56	5.16
FAN-OUT	1.89e-04	3.06e-04	97.19	5.16
FAN-IN-OUT	3.49e-04	3.82e-04	95.31	5.69
WP	5.43e-04	3.9e-04	94.06	5.16

6.10.2 Truncated weights and weight updates

In this experiment a 5-6-1 architecture was used. The weights and their updates were truncated as described in Section 6.9.2. Ten trials were performed and Table 6.16 shows the training performance. Here we see that training of the Parity 4 problem becomes much harder with CPRS and SWNP training to 100% only once. Of all five algorithms, only FAN-IN-OUT and WP managed to bring the TMSE below 0.00008. All others, managed to train to

Table 6.16 *Summary of the 10 training trials on the Parity 4 problem with truncated weights and weight updates. The update strategy PONLINE was used.*

Algorithm	Training Error		Converged
Max Iter: 4000 CEV: 0.00008	Average	Standard Deviation	
CPRS	4.42e-04	0	1
SWNP	6.4e-04	0	1
FAN-OUT	6.71e-04	3.6e-04	4
FAN-IN-OUT	2.91e-04	3.17e-04	9
WP	5.4e-04	4.1e-04	3

Table 6.17 *Computational cost the 10 training trials on the Parity 4 problem with truncated weights and weight updates. The update strategy PONLINE was used.*

Algorithm	Epochs		Feed-Forwards	
Max Iter: 4000 CEV: 0.00008	Average	Standard Deviation	Average	Standard Deviation
CPRS	4000	0	8000	0
SWNP	4000	0	32000	0
FAN-OUT	4000	0	48000	0
FAN-IN-OUT	3204	1278	22428	8946
WP	3432	983	126984	36371

100% for the number of trials shown in the 'converged' column of Table 6.16. We show in Table 6.17 the computational cost of each of the algorithms and in Table 6.18 the training performance for all trials (regardless of whether they converged or not). This table shows that FAN-IN-OUT was the most successful on average and the most consistent, followed by FAN-OUT.

Table 6.18 *Training performance with truncated weights and weight updates on all trials for the Parity 4 problem (including trials that did not converge).*

Algorithm	Testing Error		Classification (% correct)	
	Average	Standard Deviation	Average	Standard Deviation
CPRS	2.12e-03	1.5e-03	70.62	20.42
SWNP	1.23e-03	1.06e-03	81.25	18.87
FAN-OUT	3.81e-04	3.66e-04	94.38	5.47
FAN-IN-OUT	2.54e-05	8.03e-05	99.38	1.98
WP	1.98e-03	2.25e-03	83.12	16.15

6.11 Discussion

The experiments above indicate that:

- all algorithms used here can train AMLPs in an in-loop mode.

- semi-parallel weight perturbation algorithms can be much more effective than fully parallel perturbation schemes.

- when weights and weight updates are truncated, the performance of the back-propagation algorithm rapidly degrades as it has been reported in [Xie and Jabri (1992a)] and [Xie and Jabri (1992b)].

- FAN-IN-OUT has consistently produced the best performance when weights and weight updates are truncated.

Of most interest to us from these experiments is the indication for on-chip training requirements. From the foregoing experiments, the on-line training strategy is the most likely candidate for on-chip learning. As mentioned earlier, batch strategies (PBATCH or CIBATCH) are not practical for perturbation based algorithms due to the additional storage requirements and high computational cost per learning iteration. The second and more important requirement is the fact that in on-chip learning, weights and weight updates have to be in fixed point quantised format. Depending on how the weights are stored (digital or analog storage) precision may vary. The bottom line is that for a highly compact implementation, it is not attractive to have high precision weight and weight update storage. In that regard, the experiments demonstrate that with the precision of the weights reduced down to 6-bits, excellent training and generalization performance can still be achieved with semi-parallel perturbation techniques.

Due to the excessive length of time required for each experiment, it was not possible to either increase the number of trials, the maximum number of iterations or the number of classification problems. Each experiment set required one to ten days and in many cases an experiment set had to be restarted because of a computer crash. Further engineering work on the interface may facilitate faster experimentations in the future.

Additionally, theses experiments were limited to algorithms that can operate on the chip and within the interface framework. There are other training algorithms, like the MADALINE Rule III or neuron perturbation [Widrow and Lehr (1990)] which could not be tested because they require write-access to the neuron states which is not possible with Kakadu's neuron design.

6.12 Conclusions

The practical capabilities of perturbation and back-propagation based algorithms on the training of an analog MLP chip were evaluated. The experiments show that all algorithms can be used successfully for in-loop training, however back-propagation algorithms have difficulties when weights and weight updates are maintained in a fixed point format (truncated). Perturbation based algorithms that rely on an averaged gradient descent perform well in practice. As expected, sequential weight perturbation is consistent in achieving good training and generalization performance but is susceptible to being trapped in local minima. The experiments demonstrate that parallel weight perturbation techniques that perform stochastic error descent (CPRS, SWNP) all achieve good generalization performance and are less susceptible to be trapped in local minima. The experiments demonstrate, that at least for the type of learning problems that have been addressed, full parallel perturbation schemes are not as effective as semi-parallel schemes such as SWNP, FAN-OUT and FAN-IN-OUT.

CHAPTER 7

A micropower intracardiac electrogram classifier

Current Implantable Cardioverter Defibrillators (ICDs) use timing based decision trees for cardiac arrhythmia classification. Timing alone does not distinguish all rhythms for all patients. Hence, more computationally intensive morphology analysis is required for complete diagnosis. The analog VLSI neural network described in this chapter has been designed to perform cardiac morphology classification tasks. Analog techniques were chosen to meet the strict power and area requirements of the implantable system while incurring the design difficulties of noise, drift and offsets inherent in analog approaches. However the robustness of the neural network architecture overcomes, to a large extent, these inherent shortcomings of the analog approach. The network is a 10:6:3 multi layer perceptron with on-chip digital weight storage. The chip also includes a bucket brigade input to feed the Intracardiac Electrogram (ICEG) to the network and a Winner Take All circuit for converting classifications to a binary representation. The the network was trained in-loop and included a commercial implantable defibrillator in the signal processing path. The system has successfully distinguished two arrhythmia classes on a morphological basis for seven different patients with an average of 95% true positive and 97% true negative detections for the dangerous arrhythm. The chip was implemented in $1.2 \mu m$ CMOS and consumes less than $200 nW$ maximum average power from a $3V$ supply in an area of $2.2 \times 2.2 mm^2$.

7.1 Introduction

Due to improvements in battery technology, high energy circuits and packaging technology, Implantable Cardioverter Defibrillators (ICDs) are able to administer a wide variety of pacing and shock therapies to patients suffering from potentially fatal cardiac arrhythmias. This includes $600V$, $40J$ shocks for cases of Ventricular Fibrillation (VF) and certain cases of Ventricular Tachycardia (VT). With the increasing range of sophisticated and powerful therapies available, the correct classification of arrhythmia by the implant's automatic classification system is all the more important, to avoid unnecessary administering of potentially painful therapies.

To the present time, most implantable systems have used timing information from ventricular leads only to classify rhythms. This limited use of the signal from the heart's ventricles has meant that some dangerous rhythms cannot be distinguished from safe ones. By using timing information from atrial leads a larger class of rhythms may be classified via ventricular/atrial differential timing features. However, even two lead systems fail to distinguish some rhythms when timing information alone is used [Leong and Jabri (1992a)]. A case in point is the separation of Sinus Tachycardia (ST) from Ventricular Tachycardia with 1:1 retrograde conduction. ST is a safe arrhythmia which may occur during vigorous exercise and is characterized by a heart rate of approximately 120 beats/minute. VT retrograde 1:1 is also characterized by a rate of 120 beats/minute but can be a potentially fatal condition. Figure 7.1 shows the difference in the morphologies, on the ventricular lead, for these rhythms. Note, that the morphology change is predominantly in the 'QRS complex' where the letters QRS are the conventional labels for the different points in the conduction cycle. The QRS complex is the shorter window of the total beat period during which the heart is actually pumping blood.

For a number of years, researchers have studied template matching schemes in order to try and detect such morphology changes. However, techniques such as correlation waveform analysis (CWA) [Lin, DiCarlo and Jenkins (1988)], though quite successful, are too computationally intensive to meet power requirements, especially when implemented in software on the implant's microprocessor. In this chapter, we demonstrate that an analog VLSI neural network can detect such morphology changes while still meeting the strict power and area requirements of an implantable system. The advantages of an analog approach are borne out when one considers that an energy efficient analog to digital converter such as [Kusumoto and et. al. (1993)] uses $1.5nJ$ per conversion implying $375nW$ power consumption for analog to digital conversion of the ICEG alone. Hence, the integration of a bucket brigade device and analog neural network provides a very efficient way of interfac-

amplitude

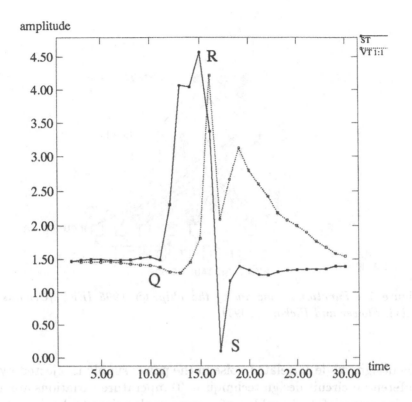

Figure 7.1 *The morphology of ST and VT retrograde 1:1.* © *1995 IEEE [Coggins, Jabri, Flower and Pickard (1995)].*

ing to the analog domain. Additionally, the use of differential pair multipliers and current node summing in the network allows a minimum of devices in the network itself and hence associated savings in power and area. However, in the last few decades analog signal processing has been used sparingly due to the effects of device offsets, noise and drift*. The neural network architecture alleviates these problems to a large extent due to the fact that it is both highly parallel and adaptive. The fact that the network is trained to recognize morphologies with the analog circuits in-loop means that the synaptic weights can be adapted to cancel device offsets [Castro, Tam and Holler (1993); Castro and Sweet (1993)]. The impact of local uncorrelated noise is reduced by the parallelism of

---

* Most fabrication processes have been optimised for digital design techniques which results in poor analog performance.

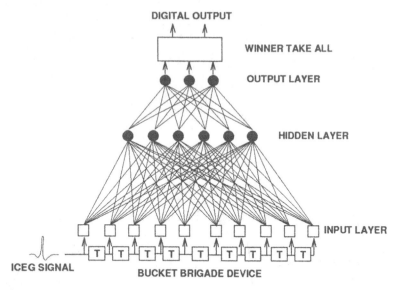

Figure 7.2 *Functional diagram of the chip.* © *1995 IEEE [Coggins, Jabri, Flower and Pickard (1995)].*

the design, while correlated noise in the power supply is rejected by differential circuit design techniques. Temperature variations are a major source of drift problems for some analog circuits. In the case of implantable systems, temperature compensation is not required due to the stable thermal environment of the human body.

The next section discusses the circuit designs in detail. Section 7.3 describes the method for training the network for the morphology classification task. Section 7.4 describes the classifier performance on seven patients with arrhythmia which cannot be distinguished using the heart rate only. Section 7.5 summarizes the results, remaining problems and future directions for the work.

7.2 Architecture

The neural network chip consists of a 10:6:3 multilayer perceptron, an input bucket brigade device (BBD) and a winner take all circuit at the output. A functional diagram of the network is shown in Figure 7.2. In this figure, circles represent neurons and the lines represent synapses. See Section 2.5 for an introduction to the multilayer perceptron. The function of the multilayer perceptron can be described mathematically by considering the output of a neuron in layer $l + 1$ given by,

$$a_i(l + 1) = f(\sum_{j=1}^{N_l} w_{ij} a_j(l)) \qquad (7.1)$$

where N_l is the number of neurons in layer l, i is the neuron index $(1 \leq i \leq N_l)$, w_{ij} is the weight connecting neuron j in layer l to neuron i in layer $l + 1$ and f is the neuron squashing function. For this design, the squashing function is implemented in a distributed manner at the input of the following layer of synapses rather than conventionally at the output of the neuron. The synapse design here, when biased in weak inversion, yields $f(x) = tanh(x)$, a commonly used squashing function. This is discussed further in Section 7.2.2.

A floor plan of the chip appears in Figure 7.3 and a photomicrograph in Figure 7.4. The BBD samples the incoming ICEG at a rate of $250Hz$. The first layer of synapses and corresponding weights determine hyperplanes which are then combined by the second layer of synapses and weights to form decision surfaces in the 10 dimensional input space. For three class problems the winner take all circuit converts the winning class to a digital signal. For the two class problem considered in this paper a simple thresholding function suffices. The following sections briefly describe the functional elements of the chip.

7.2.1 Bucket brigade device (BBD)

One stage of the bucket brigade circuit is shown in Figure 7.5. The BBD uses a two phase clock to shift charge from cell to cell and is based on a design by Leong [Leong (1992a)] which in turn is modelled on circuits developed in the early 1970s. A good survey of BBD designs also appears in Henderson (1990). The BBD operates by transferring charge deficits from S to D in each of the cells. ϕ_1 and ϕ_2 are two phase non-overlapping clocks. Assume a charge deficit is stored at node S and the voltage at D is $V_{DD} - V_T$. As ϕ_2 is brought to V_{DD} the voltage at node D goes to $2V_{DD} - V_T$ and then the charge deficit at S flows to D and the node S reaches a voltage of $V_{DD} - V_T$ whereas at node D, when ϕ_2 goes to ground again, the voltage is $V_{DD} - V_T - \Delta V$, where ΔV corresponds to

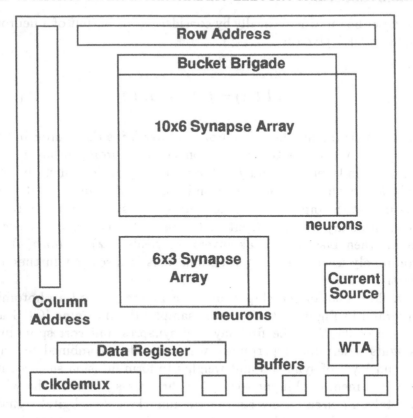

Figure 7.3 *Floor plan of the chip.* © *1995 IEEE [Coggins, Jabri, Flower and Pickard (1995)].*

the charge deficit transferred. Similarly, a new charge deficit is transferred to the input side of the cell when ϕ_1 is active.

Measurements show the unloaded BBD to have a low frequency charge transfer inefficiency of 0.35% [Leong (1992a)]. The cell is buffered from the synapse array to maintain high charge transfer efficiency. A sample and hold facility is provided to store the input on the gates of the synapses. The BBD clocks are generated off chip and are controlled by the QRS complex detector. When operated at a rate of $250Hz$ the BBD consumes $40nW$ power.

7.2.2 Synapse

The synapse circuit used is the weighted current MDAC described in Section 4.2.1 and is shown in Figure 7.7. This synapse has al-

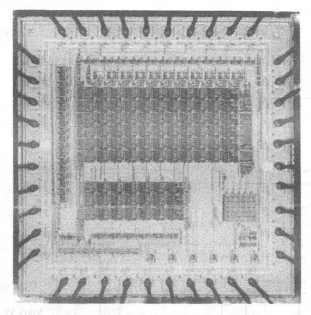

Figure 7.4 *Photomicrograph of the chip.* © *1995 IEEE [Coggins, Jabri, Flower and Pickard (1995)].*

ready been used on a number of neural network chips, [Leong and Jabri (1993a); Coggins and Jabri (1994); Pickard, Jabri and Flower (1993)]. The synapse has five bits plus sign weight storage which sets the bias to a differential pair which performs the multiplication. The bias reference is derived from a weighted current source in the corner of the chip, (See Section 4.2.2). For the neuron used in conjunction with the synapse on this chip a value of $75nA$ was chosen for the bias current to set the neuron gain (a function of the common mode current) to a suitable range for training.

Assuming the transistors forming the differential pair in the synapse are in weak inversion, then the transfer function (as shown in Section 4.2.1) of the synapse is given by

$$I_{out+} - I_{out-} = \begin{cases} +I_{DAC} \tanh(\frac{\kappa(V_+ - V_-)}{2}) & \text{if B5} = 1 \\ -I_{DAC} \tanh(\frac{\kappa(V_+ - V_-)}{2}) & \text{if B5} = 0 \end{cases} \quad (7.2)$$

where κ depends on the slope factor and the thermal voltage. Due to the integrated architecture of the chip, the synapse transfer function cannot be directly measured independent of the in-

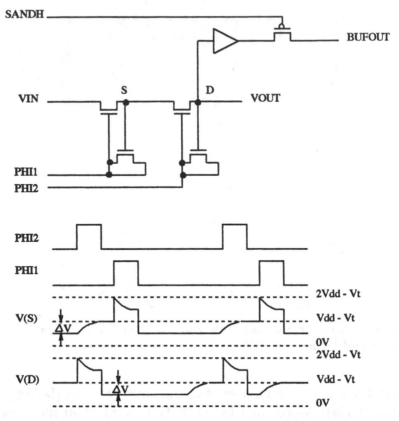

Figure 7.5 *Bucket brigade cell circuit diagram.* © *1995 IEEE [Coggins, Jabri, Flower and Pickard (1995)].*

put bucket brigade and the neuron characteristics. However, the synapse was extensively characterized on a previous chip [Coggins and Jabri (1994)] and the measured synapse transfer function from this earlier implementation is shown in Figure 7.6. The figure shows the differential input voltage $V_+ - V_-$ versus differential output voltage (linearly related to the synapse differential output current $I_{out+} - I_{out-}$) for a number of different digital weight settings of different signs. Note, the nonlinearity determined by the voltage input to the synapse. It is this nonlinearity which determines the network nonlinearity, rather than the conventional approach of implementing nonlinearity at the output of the neuron. That is, the squashing function is distributed across the proceeding layer of synapses. This can be seen to be functionally equivalent to the conventional case of using neuron non-linearity, provided that the

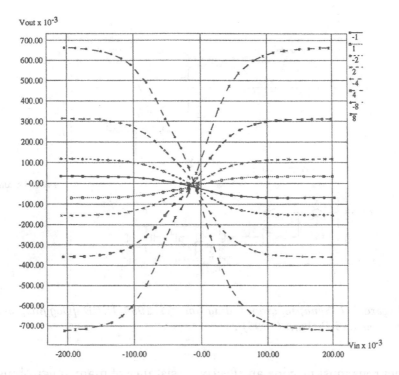

Vout x 10^{-3}

Vin x 10^{-3}

Figure 7.6 *Measured synapse transfer function for a number of differ-ent weight settings.* © *1995 IEEE [Coggins, Jabri, Flower and Pickard (1995)].*

inputs to the network correspond to the linear range of the first layer of synapses and subsequent layers of synapses have mutu-ally identical transfer functions. The first requirement is achieved by appropriate input voltage scaling, however the second require-ment is only approximately true due to device mismatch. In-loop training of the chip overcomes this effect by adapting the weights.

7.2.3 Neuron

The neuron circuit used is the common mode cancelling neuron described in Section 4.2.4 and is shown in Figure 7.8. This neu-ron circuit has been previously tested [Pickard, Jabri and Flower (1993)]. Due to the low power requirements, the bias currents of the synapse arrays are of the order of hundreds of nA, hence the

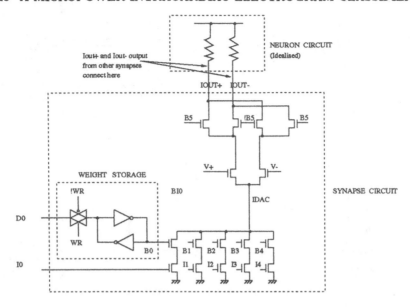

Figure 7.7 *Synapse circuit diagram.* © *1995 IEEE [Coggins, Jabri, Flower and Pickard (1995)].*

neurons must provide an effective resistance of many mega ohms to feed the next synapse layer while also providing gain control. Without special high resistance polysilicon, simple resistive neurons use prohibitively large area. A number of neuron designs have been investigated previously to solve these problems [Coggins, Jabri and Pickard (1993)]. However, for larger networks with fan-in much greater than ten, an additional problem of common mode cancellation is encountered. That is, as the fan-in increases, a larger common mode range is required or a cancellation scheme using common mode feedback is needed [Castro, Tam and Holler (1993)].

As derived in Section 4.2.4 the small signal equivalent resistance (gain) of the neuron is given by

$$R_{eq} = \frac{1}{2\sqrt{I_{cm}/(2\alpha + 1)\beta}}$$

$$= \sqrt{\frac{\alpha(2\alpha + 1)}{4I_{cm}k_p}} \qquad (7.3)$$

where $\beta = W_{0,1}/L_{0,1}k_p$ and substituting $\alpha = L_{0,1}/W_{0,1}$.

The implemented neuron uses a 3 bit SRAM to provide 8 pro-

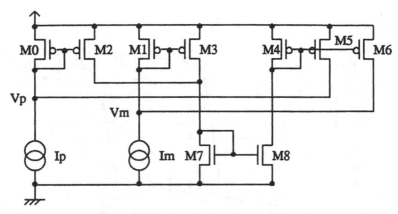

Figure 7.8 *Neuron circuit diagram.* © *1995 IEEE [Coggins, Jabri, Flower and Pickard (1995)].*

grammable gains by switching binary combinations of 3 different aspect ratio mirror transistors. The measured neuron characteristics are shown in Figure 7.9. The neuron characteristics correspond to various values of α where the 'high z' curve corresponds to M_0 and M_1 open circuited (see Figure 7.8). Here, the voltage inputs to 10 synapses with unity weights were swept. The least significant bit current was $1.8nA$.

Referring to Figure 7.6 it can be seen that over the input voltage range the neuron was tested on, the synapses were operating in their linear region and hence the nonlinearity of Figure 7.9 is due to the neuron. The sharp knee points occur as one of the diodes stops conducting significant current. This knee point is not detrimental to the training of the network, provided α is chosen such that the linear range of the neuron exceeds the saturation limits of the synapses in the following layer. Unfortunately, the nonlinear output range of the neuron also corresponds to slow operation, since the output node driving the next layer has high impedance. This should not be a problem during training, as the optimization process will tend to keep the neuron out of the nonlinear region, since it corresponds to the input saturation range of the synapse, (this will only be true if α is small enough for the linear output range of the neuron to cover the synapse input saturation range). A discussion of how chip in-loop training can overcome network imperfections appears in [Tam, Gupta, Castro and Holler (1990)]. However, when generalizing, some unexpectedly long delays may

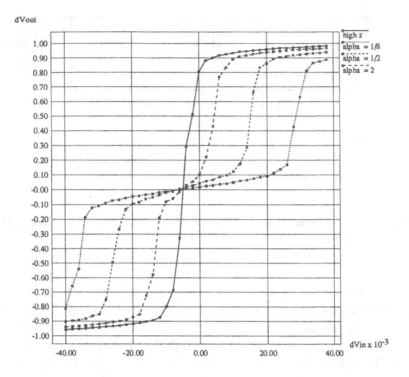

Figure 7.9 *Measured neuron transfer function.* © *1995 IEEE [Coggins, Jabri, Flower and Pickard (1995)].*

occur should an unknown pattern drive some of the neurons into the nonlinear region. In the case of ICEG classification, the bias to the weighted current source feeding the synapse arrays and the buffer bias are duty cycled on each heart beat, so the time taken for the neuron to come out of the nonlinear region is actually dominated by the bias settling time.

7.3 Training system

The system used to train and test the neural network is shown in Figure 7.10. Control of training and testing takes place on the PC. The PC uses a PC-LAB card to provide analog and digital I/O. The PC plays the ICEG signal to the input of the commercial ICD in real time. The commercial ICD performs the functions of bandpass filtering and QRS complex detection. The QRS complex

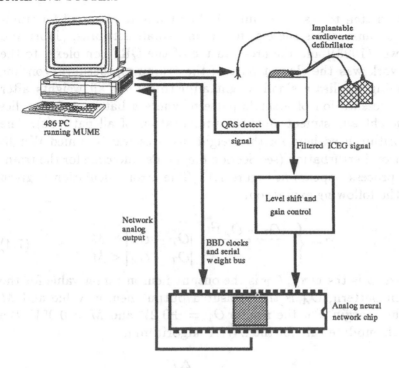

Figure 7.10 *Block diagram of the training and testing system.*

detection signal is then used to freeze the BBD clocks of the chip, so that a classification can take place.

The system has two modes of operation, namely training and generalization. In training mode, a number of examples of the arrhythmia to be classified are selected from a single patient data base recorded during an electrophysiological study and previously classified by a cardiologist. Since most of the morphological information is in the QRS complex only these segments of the data are repeatedly presented to the network. The weights are adjusted according to the training algorithm running on the PC using the analog outputs of the network to determine the weight adjustment to be made and thereby reduce the output error. The PC writes weights to the chip via the digital I/Os of the PC-LAB card and the serial weight bus of the network. The software package implementing the training and testing, called MUME [Jabri, Tinker and Leerink (1994)], provides a suite of training algorithms and control options. A special backend module for the hardware interface was written to train the chip. On-line training was used for training because of its success in training small networks, [Jabri and Flower (1993)] and the presentation of the QRS complexes to the network was the slowest part of the training procedure (on-line training implies a small weight adjustment for all weights after the presentation of a single pattern whereas batch mode implies a weight adjustment after the presentation of all patterns). The algorithm used here for the weight updates was Summed Weight Neuron Perturbation (see Section 6.5.4). Pseudo code for the training process appears in Figure 7.11. The error calculation is given by the following expression,

$$E = \begin{cases} \frac{(O_T - O_A)^2}{2} & |O_T - O_A| > M \\ 0 & |O_T - O_A| \le M \end{cases} \qquad (7.4)$$

where E is the error, O_T is the output neuron target value for the given pattern, O_A is the measured output neuron value and M is the margin. For the results $O_T = \pm 0.2V$ and $M = 0.05V$, the weight update rule for the SWNP algorithm is

$$\Delta w = -\eta \frac{\Delta E}{\gamma} \qquad (7.5)$$

where Δw is the weight update, η is the learning rate, γ is the weight perturbation and ΔE is the resulting error difference due to the perturbation.

The training proceeds in real time as the PC plays back the digitized training patterns and they are filtered and QRS detected by the ICD before being applied to the input to the network. Note, that the training patterns are subject to additional noise and distortion as they are processed by the ICD and they are also subject to the phase noise of the ICD QRS detector when they are frozen in the BBD at the input of the network. Additionally, the analog synapses and neurons contribute noise and offsets as they perform the classification. Since all these effects are taking place in the training loop in real time, they constitute additional constraints on the training problem. Hence, the weight sets obtained for the classification when the training algorithm has converged are more robust to the noise and offsets incurred in the analog signal processing, than idealized training models in numerical simulations. Furthermore, such an approach to training, whereby the analog signal processing elements are present in the training loop,

```
while  (total error > error threshold)
{
        for  (all patterns in training set)
        {
                select next training pattern
                forward propagate()
                calculate error
                accumulate total error
                for  (all non input neurons)
                {
                        for  (all weights of current neuron)
                                apply perturbation
                        forward propagate()
                        calculate error and error difference
                        for (all weights of current neuron)
                                update weight
                        forward propagate()
                        calculate error
                }
        }
}

forward propagate()
{
        do
        {
                write ICEG sample to ICD
                clock BBD
        }
        until (QRS complex detected by ICD and centered)
        measure network output
}
```

Figure 7.11 *Pseudo code for morphology classifier training procedure.
The training algorithm is SWNP using the PONLINE strategy (see Section 6.4).* © *1995 IEEE [Coggins, Jabri, Flower and Pickard (1995)].*

allows the problems likely to be encountered when training is implemented on-chip to be investigated. Continuous adaptation of
the classifier will be an important capability since over a period of
time morphology can change as a result of drug therapy and tissue
growth.

An example of a training run is depicted in Figure 7.12. The figure shows the training error after each set of training patterns has
been applied and the corresponding weight updates performed. Notice, that as the error approaches lower values, the error becomes
more noisy. Two reasons for this are the fixed step size of the optimization algorithm and the noise in the analog signal processing.
Table 7.1 summarizes the training statistics for the seven patients.
Notice that 100% training performance is obtained for all but two
of the patients. (Each of the patients was trained with eight examples of each class of arrhythmia).

Once the morphologies to be distinguished have been learned

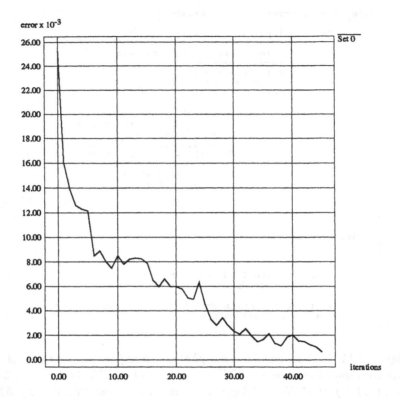

Figure 7.12 *Graph of training error versus iterations.* © *1995 IEEE [Coggins, Jabri, Flower and Pickard (1995)].*

Table 7.1 *Training performance of the network on seven patients with ICD in-loop.* © *1995 IEEE [Coggins, Jabri, Flower and Pickard (1995)].*

Patient	Training Iterations	% Correct	
		ST	VT
1	56	100	100
2	200+	100	87.5
3	200+	87.5	100
4	46	100	100
5	200+	100	100
6	140	100	100
7	14	100	100

Table 7.2 *Classification performance of the network on seven patients with ICD in loop.* © 1995 IEEE [Coggins, Jabri, Flower and Pickard (1995)].

Patient	No. of complexes		% Correct	
	ST	VT	ST	VT
1	440	61	100	98.3
2	94	57	100	95
3	67	146	77.6	99.3
4	166	65	91	99.3
5	61	96	97	93
6	61	99	97	100
7	28	80	96	99

for a given patient, the remainder of the patient database is played back in a continuous stream and the outputs of the classifier at each QRS complex are logged and may be compared to the classifications of a cardiologist. The ability of the network to classify previously unseen patterns is referred to as its generalization performance and is discussed in the next section.

7.4 Classification performance and power consumption

The system was trained on seven different patients separately, all of whom had VT with one to one retrograde conduction. Note, that patient independent training has been tried but with mixed results [Tinker (1992)]. Table 7.2 summarizes the performance of the system on the seven patients. Most of the patients show a correct classification rate well over 90%, whereas, a timing based classifier cannot separate these arrhythmia. These results may be compared to those found in Throne, Jenkins and DiCarlo (1991), where a number of digital template matching techniques are evaluated for sinus rhythm and ventricular tachycardia separation and computational complexity. In this study the Normalized Area Difference method achieved best results with 100% separation for the 19 patients considered.

The network architecture used to obtain these results was 10:6:1, the unused neurons being disabled by setting their input weights to zero. The input signal was sampled at a rate of $250Hz$. The power

Table 7.3 *Power consumption of the chip.* © 1995 IEEE [Coggins, Jabri, Flower and Pickard (1995)].

Module	Power (nW)	Conditions
BBD	40	250Hz rate
Network	128	75nA LSB
Buffers	18	3uA bias
Total	186	120bpm

supply was $3V$. With a nominal heart rate of 120 beats/minute and duty cycling the bias to the synapse arrays and buffers with a bias settling time of $70\mu s$ the power consumption figures for the chip are shown in Table 7.3.

7.5 Discussion

7.5.1 Results

Referring to Table 7.1 we see that patients 2 and 3 were not trained to perfectly distinguish the classes. In the case of patient 2 this did not appear to affect the generalization performance significantly. Hence, this may be due to a rare VT morphology or more likely, the training algorithm being trapped in a local minima. However, for the case of patient 3, a high false positive error rate for VT is obtained (i.e. many of the STs were classified as VT). Inspection of this patient's data showed that the morphologies of the two rhythms were very close. A typical example of this is shown in Figure 7.13. The figure shows that the width of the QRS complex has not changed and contrasts sharply with that of Figure 7.1.

7.5.2 Improving the classification

By varying the form of the error function during training a number of improvements could be achieved in the classification performance. This would include biasing the error function to reduce the false positive error rate i.e. to reduce the chance of a dangerous arrhythmia not being treated. Using hints as described in Abu-Mostafa (1993) during training to reduce the classifier sensi-

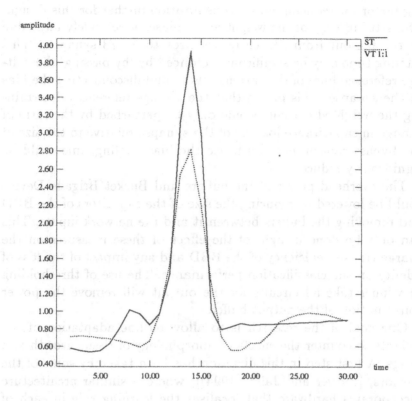

amplitude

ST
VT 1:1

time

Figure 7.13 *A typical morphology of ST and VT retrograde 1:1 for patient 3.* © *1995 IEEE [Coggins, Jabri, Flower and Pickard (1995)].*

tivity to amplitude, shift and DC variations may assist extending the classification to a more patient independent basis and reduce sensitivity to morphology change over time for a single patient.

More work is needed to determine the optimum trade-offs between weight precision, gain precision and the rate of convergence for training. These factors in turn influence the power and area used by the network and feasibility of implementing learning algorithms on chip.

7.5.3 Improving circuit performance

The slow heart rate relative to the propagation time of the network means that the best power consumption is obtained by duty cycling the bias currents to the synapses and the buffers. Hence, the bias settling time of the weighted current source is the limiting factor for reducing power consumption further for this design. The settling time of the weight array biases is relatively large due to the fan-out from one current source to all 78 synapses. This settling time may be significantly reduced by 'by passing' the voltage reference lines of the current source and disconnecting the bias at the synapse. This means that the voltage references determining the weighted currents would only be perturbed by the ratio of change in capacitance loading of the synapse relative to the size of the bypass capacitance and hence the bias settling time could be significantly reduced.

The overhead power of the buffers and Bucket Brigade Device could be lowered by reducing the size of the capacitors of the BBD and removing the buffers between it and the network inputs. This can only be done in light of the effect of these measures on the charge transfer efficiency of the BBD and any impact of the loss of fidelity on the classification performance. The use of thresholding or winner take all circuits for the output will remove the power consumption of the output buffers.

One goal of the research is to allow on-line adaptation of the weights to counter the effects on morphology of tissue growth and drugs. A first step in this direction has been taken by some of the authors, [Flower and Jabri (1994)], where a similar architecture incorporates hardware that localises the learning rule in each of the synapses.

7.6 Conclusion

The successful classification of a difficult cardiac arrhythmia problem has been demonstrated using an analog VLSI neural network approach. Furthermore, the chip developed has shown very low power consumption of less than $200nW$, meeting the requirements of an implantable system. The chip has performed well, with over 90% classification performance for most patients studied and has proved to be robust when the real world conditions of QRS detection jitter, noise and distortion are introduced by a commercial implantable cardioverter defibrillator placed in the signal path to the classifier.

CHAPTER 8

On-chip perturbation based learning

8.1 Introduction

Approximate gradient descent algorithms such as Weight Perturbation (WP) [Jabri and Flower (1992)], Node Perturbation (NP) or Madeline Rule III (MRIII) [Widrow and Lehr (1990)] and stochastic gradient descent algorithms such as Summed Weight Neuron Perturbation (SWNP) [Flower and Jabri (1993a)], Stochastic Error Descent (SED) [Cauwenberghs (1993)] or Parallel Gradient Descent [Alspector, Meir, Yuhas, Jayakumar, Lippe (1992)], belong to a class of artificial neural network training algorithms that are particularly well suited to implementation in VLSI. The two chips described in this case study are of different architectures, one is a Multi Layer Perceptron (MLP) while the other a is Recurrent Network, however they use similar training algorithms. These algorithms are computationally more complex than Backpropagation (BP), but they are easier to implement in VLSI as they were developed specifically for this purpose. They address several issues that have hampered the implementation of the exact gradient descent algorithms. These include scalability, precision requirements, neuron activation function derivative generation, drift, noise and cost in terms of hardware dedicated to the training procedure.

8.2 On-chip learning multi-layer perceptron

The design and implementation of a chip called Tarana is discussed in this section.

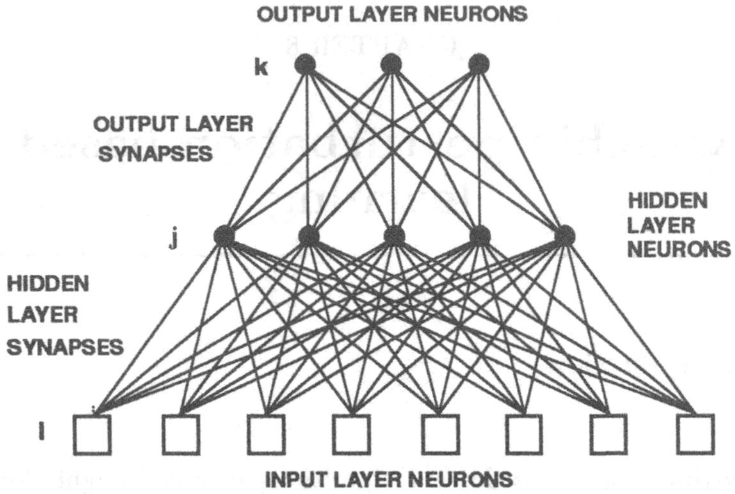

Figure 8.1 *Schematic representation of the architecture of the Tarana chip. The chip contains 8 hidden and output neurons with a hyperbolic tangent-like transfer function The input neurons are linear with a gain of one. There are 55 synapses; 40 in the hidden layer and 15 in the output layer.*

8.2.1 Network architecture

The MLP implemented on Tarana has three layers, consisting of 8 inputs, 5 hidden layer neurons and three output layer neurons. A total of 8 neurons and 55 synapses (see Figure 8.1).

The signalling in Tarana is differential, which delivers better noise immunity than single ended signals at the cost of area. The synaptic input is a pair of differential voltages. The synaptic output is a pair of differential currents that may be summed* with the outputs of other synapses before being applied to the neuron. The neuron behaves like a passive non-linear resistor converting the current outputs of the synapses to a differential pair of voltages. The synapses and neurons are individually and group addressable (as a row) to enable the weights or gains to be set and read, the perturbation bit to be set, and a perturbation to be applied. The MLP architecture implements the feedforward dynamics,

---

* This summation simply exploits Kirchhoff's current law.

$$net_j = \sum_{i \in L_j} w_{ij} x_i \tag{8.1}$$

where $x_i = f_i(net_i)$, w_{ij} is the weight of the synapse connecting the output of the ith neuron to the input of the jth neuron, x_i is the output of the ith neuron, L_j is the set of neurons connected to the jth neuron, net_i is the sum of the synapse outputs leading into the ith neuron, and $f_i()$ is the transfer function of the ith neuron.

The value of the synaptic weight may be modified using one of two methods. Either an analog voltage may be written directly to the weight or the charge on the storage capacitor may be increased or reduced by means of a current source circuit or a current sink circuit.

8.2.2 Network structure

The structure of the Tarana chip is shown via the floor plan in Figure 8.2. There are two synaptic matrices corresponding to the hidden layer synapses and the output layer synapses. The neurons form a column on the right-hand side of a synapse matrix. Other circuits implement the row/column addressing scheme, control signal demultiplexing, weight read/refresh mechanism and output signal buffering.

8.2.3 Synapse learning circuitry

The Tarana chip is implemented using a $1.2\mu m$ n-well double metal, double polysilicon fabrication process, on a die $2.2mm \times 2.2mm$ in area.

The synapse (see Figure 8.3) uses a Gilbert multiplier to provide four quadrant multiplication of the differential input voltage with a weight value stored as a voltage on a capacitor. The weight is a differential value but one of the pair of signals is kept at a fixed voltage and is connected to a common signal to reduce the I/O pin requirements. The voltage on the storage capacitor is the signal $VWEIGHT$ (i.e., the synaptic weight). This can be directly set to a voltage or it can be incrementally modified by the charge pump and sink circuitry. Additional circuitry provides the row/column address decoding and logic to control the weight perturbation and update operations. For more detail see Section 4.2.5.

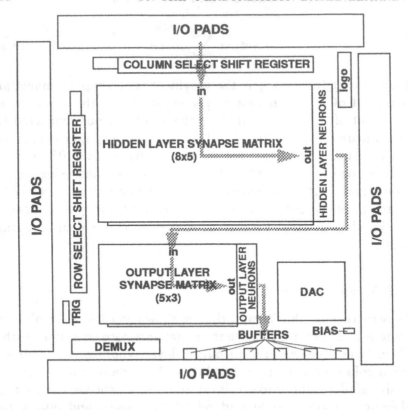

Figure 8.2 *Floor plan of Tarana chip layout showing the major physical blocks identified by function. The data flow through the network layers is indicated by the gray arrows.*

Neuron circuitry

The neuron is implemented as a Common Mode Compensating Neuron (CoMCoN) and is shown in Figure 4.5 and described in Section 4.2.4, [Pickard (1993a); Pickard (1993b)]. This is a current to voltage conversion element that has a programmable gain setting. A differential current is fed to the signals VM and VP and the effective resistance of the CoMCoN creates a differential voltage on the same signals. The common mode resistance varies inversely with the common mode current so that as the number of synapses feeding a neuron increases the neuron will maintain a constant output dynamic range.

The circuit diagram of the neuron of Figure 4.5 does not show the circuitry for variable gain ranges. This is implemented by adding

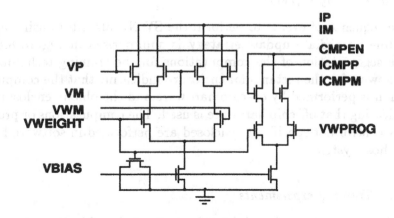

Figure 8.3 *Schematic diagram of the synapse multiplier, weight storage and comparator front end. All transistors are n-type FETs.*

PMOS FETs of varying lengths to the current mirrors directly connected to signals VP and VM.

8.2.4 Training techniques

The Tarana chip has been designed to allow the use of a range of perturbation techniques for training. This flexibility is due to two key features. Firstly, the synapses can not only be addressed individually but they can also be addressed as rows, and acted upon in parallel. This means that all the synapses that feed into a neuron can be perturbed or updated simultaneously. Secondly, the key portion of the weight update rule is implemented in hardware local to each synapse.

Weight Perturbation (WP) (see Section 6.5.2) may be performed by generating an error difference and determining its polarity. Then the selected weight is updated with a perturbation of duration proportional to the magnitude and polarity of the update required. Summed Weight Neuron Perturbation (SWNP) (Section 6.5.4) may also be performed with each row of synapses associated with a neuron perturbed and updated simultaneously.

8.2.5 Training sequence

The sequence of events to perform the SWNP algorithm using an on-line immediate update strategy is summarised in Figure 8.4. The segmentation of the computations for the training technique is shown with the rectangular enclosure indicating that the computation is performed by on-chip hardware and the oblong enclosure indicating that off-chip hardware is used. The computations or process control not specifically enclosed are performed in software by the host system.

8.2.6 Training experiments

The various subcircuits of the chip performed within their design specification and Tarana was successfully trained on several bench mark problems using a range of perturbation training techniques. The on-chip learning circuitry provided a large increase in the speed of training as it alleviated the I/O bottleneck normally associated with in-loop training and off-chip weight storage.

8.2.7 Training results for the XOR problems

The network architecture used for the XOR problem is 6:3:1. This includes 4 bias inputs and 2 inputs for the 2 bit training vector. Four bias inputs are required to provide sufficient common mode input current to the neurons. The third hidden layer neuron is needed to provide a bias input to the output layer neuron. Table 8.1 shows a summary of the results with some statistics regarding the convergence. These results consist of at least 50 trials of each of the training algorithms. The FF Ratio column indicates the number of feedforward passes performed by the controlling software for each method relative to WP. The lower the figure the faster the algorithm is in real time. OCLWP is On-Chip Learning WP and OCLSP is On-Chip Learning SWNP. AOLWP is Automatic hardware generated error measure On-Chip Learning WP and AOLSP is Automatic hardware generated error measure On-Chip Learning SWNP. Other problems have been tried on this system including the 3 bit parity problem and the 'real world' problem of ICEG morphology classification. The chip performed well, particularly for the ICEG classification because such problems benefit from the injection of noise into the system [Murray (1991)].

Initialise weights;
Start;
Set perturbation bits to random values;
Foreach input and target vector pair do
    Apply the current input and target vector pair to the network;
    Foreach neuron do
        currentMSE = MSE;
        Perturb all weights of current neuron;
        errDiff = MSE - currentMSE;
        Remove perturbation from weights;
        If the magnitude of errDiff > threshold then
            Foreach weight of current neuron do
                If errDiff * pertBit is +ve then
                    Subtract weightupdate from current weight;
                else
                    Add weightupdate to current weight;
                endif
            done
        endif
    done
done
If MSE < convergence_threshold then
    Training complete;
else
    Goto Start;
endif

Performed by on-chip hardware Performed by off-chip hardware

Figure 8.4 *Pseudo code describing the sequence of events for implementing the SWNP algorithm using the PONLINE update strategy on the Tarana chip. (see Section 6.5.4 and Section 6.4). The variable* pertBit *is the value of the current weights perturbation sign (0 = -ve, 1 = +ve),* weightupdate *is the value by which a weight is incremented or decremented and* convergence_threshold *is the MSE at which the network is considered to have learnt the problem. The enclosing boxes indicate where the various computations are performed (i.e. either in on-chip hardware or off-chip hardware). Computations and process control not enclosed are performed in software by the host.*

Table 8.1 *XOR problem training statistics. The FF ratio column is the relative number of feedforward passes that has to be performed by the controlling software. This indicates the relative speed of each of the training techniques with respect to weight perturbation.*

Training Technique	Convergence Threshold	Epochs		FF Ratio	Convergence %
		Average	SD		
WP	0.05	27.95	21.93	1.0	93.0
SWNP	0.05	59.73	40.13	0.48	76.0
OCLWP	0.05	19.09	8.79	1.33	66.0
OCLSP	0.05	58.14	38.66	0.47	92.5
AOLWP	0.05	16.16	6.32	0.05	77.0
AOLSP	0.05	109.24	78.93	0.35	74.0

Examples of the convergence trajectories for the OCLWP algorithm learning the XOR problem is shown in Figure 8.5. The average number of epochs that OCLSP requires to reach convergence is about the same as for OCLWP but the number of computations (and therefore time) is a factor of $O(N)$ less, (i.e. a factor of 3). Similar trajectories for the other algorithms, learning both the XOR and the 3 bit parity problem, were obtained, but are omitted for brevity.

8.3 On-chip learning recurrent neural network

The implementation of a Continuous Time Recurrent ANN in VLSI is illustrated using a chip design by [Cauwenberghs (1994a)]. This chip features six continuous-time analog neurons each with a variable threshold and 36 synapses with variable weights. The total of 42 free parameters may be trained using an on-chip implementation of a stochastic perturbative algorithm.

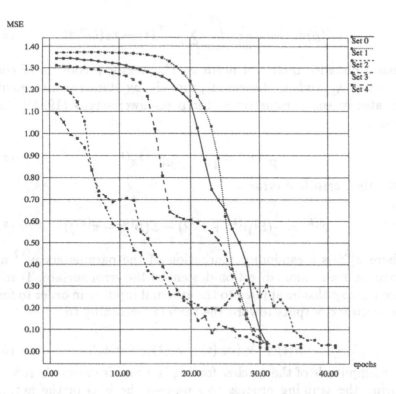

Figure 8.5 *MSE trajectories for 5 training runs using the on-chip weight update weight perturbation algorithm (OCLWP).*

8.3.1 System architecture

The architecture is fixed as a fully interconnected recurrent network with the neuron states being the outputs of the network. The dynamics of the network are described by,

$$\tau \frac{d}{dt} x_i = -x_i + \sum_{j=1}^{6} w_{ij}\, \sigma(x_j - \theta_j) + y_i , \qquad (8.2)$$

where $x_i(t)$ are the neuron states, $y_i(t)$ are the external inputs to the network and $\sigma()$ is the neurons sigmoidal activation function. w_{ij} are the synaptic weights and θ_j are the neuron thresholds. The time constant τ is fixed and identical for each neuron. The error measure is the time averaged error

$$\mathcal{E}(p) = \lim_{T \to \infty} \frac{1}{2T} \int_{-T}^{T} \sum_{k=1}^{2} |x_k^T(t) - x_k(t)|^\nu dt, \qquad (8.3)$$

with a distance metric of norm ν. The vector p consists of components w_{ij} and θ_j. Learning proceeds by iterative incremental updates in the parametric vector $\mathbf{p}$ [Cauwenberghs (1994)], such that

$$\mathbf{p}^{(k+1)} = \mathbf{p}^{(k)} - \mu \hat{\mathcal{E}}^{(k)} \pi^{(k)} \qquad (8.4)$$

with the perturbed error

$$\hat{\mathcal{E}}^{(k)} = \frac{1}{2}(\mathcal{E}(\mathbf{p}^{(k)} + \pi^{(k)}) - \mathcal{E}(\mathbf{p}^{(k)} - \pi^{(k)})) \qquad (8.5)$$

where $\pi_i^{(k)}$ is a random perturbation of the parameters $\mathbf{p}_i^{(k)}$ and learning is a random direction descent of the error surface. Teacher forcing may also be applied to the external input y_i in order to force the network outputs toward the targets, according to

$$y_i(t) = \lambda \gamma (x_i^T(t) - x_i(t)), \ i = 1, 2. \qquad (8.6)$$

The amplitude of the teacher forcing is λ and is gradually reduced during the training process to suppress the bias in the network outputs at convergence caused by residual errors.

8.3.2 Implementation

This chip was implemented in CMOS as a single chip with the majority of functions concerned with learning integrated on the chip. This excludes some global and higher level functions such as the error evaluation which allows for a greater flexibility in the learning process. The design of the neuron and synapse are briefly described in the following subsections.

8.3.3 Synapse

The synapse (see Figure 8.6) is a high impedance triode multiplier, using an adjustable regulated cascode. The current I_{ij}, which is linear with respect to W_{ij} over a wide range, feeds a differential pair, thus injecting the differential current $I_{ij}\sigma(x_j - \theta_j)$ into the diode connected output lines, I_{out}^- and I_{out}^+.

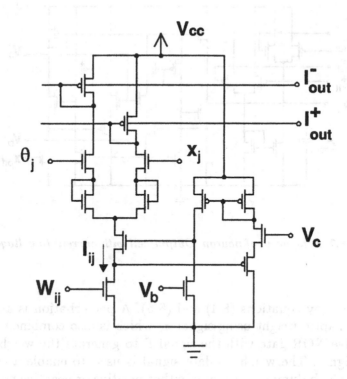

Figure 8.6 *Schematic of synapse.*

8.3.4 Neuron

The current output of a row of synapses is summed on the lines I_{out}^- and I_{out}^+ and subtracted from the reference currents I_{ref}^- and I_{ref}^+ in the neuron cell (see Figure 8.7). The output current difference is converted to the voltage x_i using regulated high output impedance triode circuitry as an active resistive element. The feedback delay τ is the RC time delay of the active resistance and the capacitor C_{out} (5 pF). This can range between 20 and 200μsec by adjusting the control voltage of the regulated cascode.

8.3.5 Learning and long term storage circuitry

The system contains specialised learning circuitry (see Figure 8.8) that is used to optimise the free parameters according to a paradigm

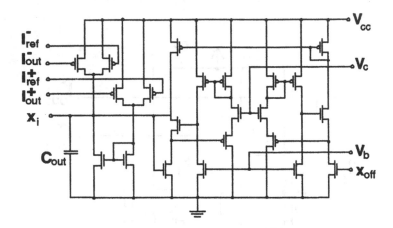

Figure 8.7 *Schematic of neuron. Output cell with current to voltage converter.*

described by equations (8.4) and (8.5). A perturbation is applied to a synaptic weight using signal π_i which is also combined in an exclusive NOR gate with the signal $\hat{\mathcal{E}}$ to generate the weight update signal. The weight update signal is used to enable a charge pump which dumps a current of either positive or negative polarity onto the storage capacitor. The circuitry to perform the training is also used in the refreshing of the analog storage values.

8.3.6 Experimental results

The recurrent network was trained to produce a circular target trajectory. This is defined by the quadrature phase oscillator,

$$
\begin{aligned}
x_1^T(t) &= A\cos(2\pi ft) \\
x_2^T(t) &= A\sin(2\pi ft)
\end{aligned}
\tag{8.7}
$$

where $A = 0.8V$ and $f = 1kHz$. This problem can be solved by two neurons but the additional neurons are required to reduce the nonlinear harmonic distortion due to the particular amplitude and frequency requirements. The network is capable of learning this problem in under 100 seconds using the on-chip learning mechanism and teacher forcing.

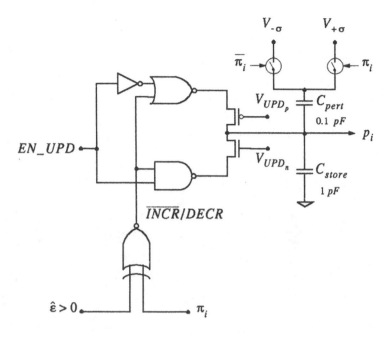

Figure 8.8 *Simplified schematic of learning cell circuitry.*

8.4 Conclusion

The two chips described in this chapter demonstrate the implementation of gradient based training algorithms in analog hardware environments. In both cases the localisation of the weight storage and weight update rule within the synapses means that the information bottleneck caused by writing weights from off-chip is avoided. This results in greatly reduced training times when compared with in-loop training methods. The different network structures and architectures represented by the Multi-Layer Perceptron and the Recurrent Neural Network are evidence of the versatility of the gradient descent based perturbation algorithms.

CHAPTER 9

An analog memory technique

Co-author: **Pascal Heim**

9.1 Introduction

The lack of convenient long-term storage for the synaptic weights is a very constraining limitation in the implementation and application of analog neural networks. Present floating-gate techniques are not convenient in this context because of their badly controlled weight update and limited number of operations.

Much work has been done to solve the problem of long-term synaptic storage, and the proposed techniques are usually best suited to a particular range of applications. The two most common alternatives to floating-gate devices are self-refreshing multi-level storage on capacitors and multiplying DACs using standard RAM cells. This chapter briefly discusses the context in which these techniques are applicable, and points out their respective advantages and limitations.

For artificial neural networks, there are basically four classes of hardware implementations which have different application potentials. By order of increasing complexity there are:

- hard-wired networks, such as silicon retinas, which do not exhibit adaptive behaviour;

- neural accelerators which are intended to increase the speed of a host computer in neural applications;

- programmable neural networks which may be trained to perform a particular task, but are not intended to autonomously adapt;

- fully adaptive neural networks which are capable of continuously learning and adapting to new environment while operating.

The self-contained storage cell described in this chapter is intended for programmable neural networks (last two classes of hardware implementations above). The cell performs both an analog to digital conversion (ADC) and a digital to analog conversion (DAC). It is capable of saving the synaptic weights in analog implementations of neural networks at the end of the training process. Its use in multi-layer perceptrons implementations with on-chip learning opens a wide range of potential applications in domains like implantable devices, consumer products and telecommunications. As an example, there are some applications in implantable devices, using small multi-layer perceptrons (see Chapters 7 and 8), which, if fabricated with a low-cost standard CMOS technology, may be used in industrial products despite the rather large area of the storage cell.

9.2 Self-refreshing storage cells

Self-refreshing storage cells are often credited with long-term storage capability. However, since they do not store information permanently, they are better classified as medium-term storage devices. The principle of these cells is based on the periodic refresh of the voltage on a capacitor to predefined quantized voltage levels. The refresh period must be short enough to compensate for current leakage associated with the access switch transistor to the storage capacitor.

One of the first self-refreshing storage cells developed for neural network implementations has been proposed by Hochet (1989), Hochet, Peiris, Abdo and Declercq (1991). The capacitor voltage is compared to a voltage ramp and the quantized voltage levels are defined by a synchronized clock, so that the synaptic weight is available both as a voltage on a capacitor and as a time delay on a digital signal. This duality has been exploited advantageously to implement the learning operations in a mixed analog-digital implementation of a Kohonen feature map [Peiris (1994)], which shows that self-refreshing storage cells have a variety of applications apart from the simple storage of information.

Another interesting version of a self-refreshing cell has been proposed by Vittoz, Oguey, Maher, Nys, Dijkstra and Chevroulet (1991), in which the quantized voltage levels are defined by a step-like reference signal common to all the cells. The capacitor voltage is directly refreshed to the closest upper step detected by a comparator. The structure is insensitive to the comparator's input offset voltage and incorporates a leakage reduction technique which can be exploited to reduce the refresh period and consequently the power consumption of the refreshing circuitry.

Following these two first refreshing schemes, various versions have been proposed [Castello, Caviglia, Franciotta and Montecchi (1991); Cauwenberghs (1994); Pedroni (1994)], the latter includes a promising error compensation principle which improves the robustness against noise. Note, however, that self-refreshing memories cannot ensure total noise immunity, which prevents their use in body-implantable device applications for example. Another drawback is the power consumed by the refresh circuitry which is not desirable in battery-powered applications. Actually, these cells are particularly well suited to implement neural accelerators. Their built-in data storage avoids the need for sequential weight refresh and, provided the weight's updates are local and analog, the parallelism of the neural network can be fully exploited.

9.3 Multiplying DACs

A second common alternative to floating-gate devices is the use of standard RAM cells to store the synaptic weights as binary values and convert them with a local DAC to produce the analog value of a weight. The effect of quantization of the weights is similar to that of self-refreshing cells, but the weight updates cannot be performed on chip. The weights must therefore be down-loaded at each learning step, which may considerably slow the training process.

This approach has been used in several neural network implementations, mostly as general purpose arrays of synapses and neurons [Raffel, Mann, Berger, Soares and Gilbert (1987); Mueller, Van der Spiegel, Blackman, Chiu, Clare, Donham, Hsieh and Loinaz (1989); Moopenn, Duong and Thakoor (1990); Alspector, Allen, Jayakumar Zeppenfeld and Meir (1991); Sackinger, Boser, Bromley, LeCun and Jackel (1992)]. In Chapters 4, and 7, multi-layer perceptrons have been implemented with $n = 5$ bits plus sign (1.5% resolution). Each bit is accessed separately to control binary-weighted current sources. Since the neural network is in the loop of the learning process, the weights' quantization and other circuit inaccuracies are compensated. However, such a scheme will suffer speed and power consumption penalties if used in on-chip learning applications. The multiplying DAC approach is therefore best suited to implement programmable neural networks that do not need to continuously learn in real-time.

9.4 A/D-D/A static storage cell

In applications where a neural network is trained and is not required to continuously adapt, its weights can be 'frozen' at the end of the learning process. Since analog neural network chips must be trained in-loop to compensate the circuit inaccuracies, the truncation of the weights at the end of the training process may decrease the average convergence rate. This can be overcome partly by restarting the training process shortly after the storage phase if the quantization of the weights results in an unacceptable increase of the mean squared error.

For on-chip learning, the interface with the host computer should be kept to a minimum, i.e. the supply of the training data base and a few learning control signals. For this purpose, the outputs of the network used during the operational phase can share the same I/O pads with the inputs used to present the teacher patterns during the training phase. This is possible only if the computation of the mean square error is made on-chip, since both the network's outputs and the teacher's inputs are needed simultaneously during the training phase. Such an approach certainly increases the chip area, but may be necessary to limit the number of pins.

On the other hand, if on-chip learning is made with short-term memories, a sequential transfer to the long-term memories may be too long compared to the decay rate of the short-term memories. The proposed storage cell performs both A/D and D/A conversions locally, and all the synaptic weights of a network can be stored in parallel at the end of the learning process.

9.5 Basic principle of the storage cell

A schematic of the basic bit-cell is depicted in Figure 9.1. The cell is made of a standard RAM cell including switch S_1, a current source I_{Bi}, four additional switches $S_2 - S_4$ and M_{14}, an input current node I_{in} and an output current node I_{out}. A complete n-bit storage cell is made of n identical cells in cascade with binary-weighted current sources I_{Bi} as shown in Figure 9.2. The complete transistor schematic of a single cell is shown in Figure 9.3.

In the restore mode ($C_i = 1$ in Figure 9.2), switches S_1 and S_4

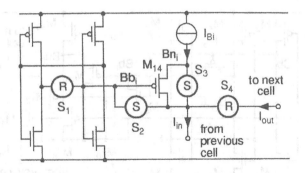

Figure 9.1 *Schematic of a single bit cell of the static storage device. The switches labelled R are turned on in the restore state, and those labelled S are turned on during the storage or A/D conversion cycle.*

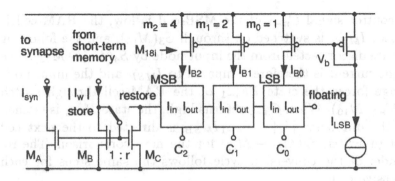

Figure 9.2 *Three-bit storage cell. The bit current sources are made of parallel combinations of m_i transistors.*

are on, and switches S_2 and S_3 are off. The RAM cells are latched and all the cells are shorted together by switch S_4. The current sources I_{Bi} are connected through transistors M_{14} depending on the contents of the RAM cells, and the stored value is available as a current source at the input node I_{in} of the MSB cell. In this mode, the bidirectional current mirror $M_B - M_C$ is in the restore position. The exact value of the mirror ratio r does not affect the scale of the synaptic current I_{syn} between the store and restore phases, because the mirror ratio is inverted during the restore phase.

The storage phase performs a successive approximation conversion. Each cell is successively accessed from the MSB to the LSB through the control signals C_i as shown in Figure 9.4. When

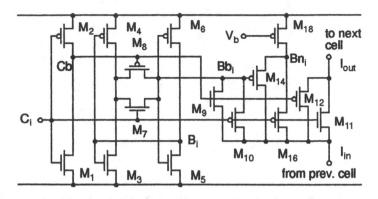

Figure 9.3 *Complete schematic of the basic bit-cell.*

the control signal C_{n-1} of the MSB-cell is low, the RAM-cell is opened, I_{MSB} is switched-on through $S_3(M_{16})$, and the following cells are disconnected from the input node by $S_4(M_{11} - M_{12})$. The weight current is therefore compared to I_{MSB} and the input node voltage forces the state Bb_{n-1} of the RAM-cell through switch $S_2(M_9 - M_{10})$. When C_{n-1} returns high, the state B_{n-1} is latched and the difference $|I_w| - B_{n-1}I_{MSB}$ is directed to the next cell through switch $S_4(M_{11} - M_{12})$ for the next comparison. The remainder of the conversion cycle follows the same steps for each successive cell.

9.6 Circuit limitations

To ensure monotonicity, the difference between any bit current and the sum of the lower order bit currents must not exceed one LSB. The required relative precision is divided by 2 for each subsequent cell. A good approach is to use weighted current sources made of parallel transistors with the same gate voltage, as shown in Figure 9.2. By this means, the total active area of the current sources is used to match the MSB current with the sum of all the other bit currents. Another source of inaccuracy is due to the limited current comparison time when the current difference is small. The control pulses must be long enough so that the maximum accumulated error after the n-bit conversion does not exceed 1/2 LSB. Since each cell is successively connected to the previous ones during the conversion, the node capacitance increases linearly and the control

Figure 9.4 *Control signal timing for the storage phase of an n-bit storage cell.*

pulses must be as shown in Figure 9.4. As an example, with a supply voltage of $3V$, $I_{LSB} = 1nA$, $C_{node} = 50fF$, the total conversion cycle needs about $10ms$ for $n = 5$, which is compatible with currently attainable decay rate of short-term memories. Finally, the current comparison is also affected by the threshold voltage mismatch between inverters $M_5 - M_6$. This source of inaccuracy can be drastically reduced by using long-channel or cascoded current sources to obtain a high impedance at the comparison nodes Bn_i. This also applies to the current mirror $M_B - M_C$, because the drain voltage of M_C, if different in the two modes, may change the synaptic current scale. If all these limitations are carefully taken into account, the accuracy of the storage cell is basically identical to that of multiplying DACs, provided the accuracies of the current sources are similar.

9.7 Layout considerations

The layout of a 5-bit storage cell has been designed using lambda-based design rules. The size of a complete 5-bit storage cell is $200\lambda \times 108\lambda$ (width $\times$ length), including space for abutting circuitry. For a $1.2\mu m$ technology, the cell area is $7776\mu m^2$, which is acceptable in various practical applications. The size of a single basic cell is $40\lambda \times 79\lambda$ and its area could be reduced by about 20 % by using full-custom design rules with 45 degree drawing rules. The weighted current sources are made respectively of 8, 4, 2, 1 unit transistors in parallel and 2 unit transistors in series for the LSB. This configuration optimizes the active-to-total area ratio and consequently the current sources matching. The current sources are cascoded and the transistors sizes are $8\lambda/10\lambda$, which is 10 times larger than the minimum device area. The accuracy of

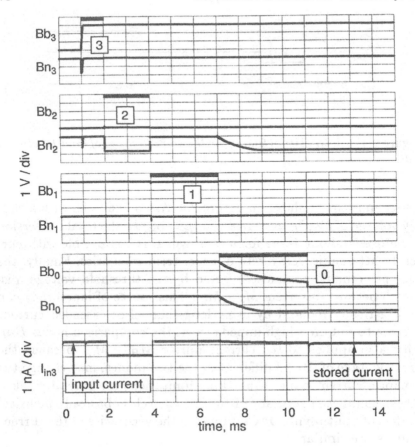

Figure 9.5 *Transient simulation of a four-bit storage cell, showing all possible bit transitions*

the MSB is that of 8 transistors in parallel and should be suitable for 5 bit resolution.

9.8 Simulation results

A complete conversion cycle of a 4-bit storage cell has been simulated. The technological parameters are those of a standard $1.2\mu m$ CMOS technology, and the parasitic capacitances have been extracted from the above-mentioned layout, which will be used for the integration of a multi-layer perceptron with on-chip learning. Figure 9.5 shows the transient simulation of a 4-bit version of this storage cell, with $I_{LSB} = 1nA$. The initial state of the cell is $B = 1100$ (or $Bb = 0011$), which represents a stored current of $12nA$. The input current to the cell before conversion is $|I_w| = 5.1nA$, which must produce the code $B = 0101$ (or $Bb = 1010$). This situation shows the four possible bit transitions. The input current I_{in3} on the figure is the drain current of M_C.

The complete conversion cycle can be described as follows:

- Cell number 3: $I_{MSB} = I_{B3}$ is larger than I_{in3}, which pulls Bb_3 high and switches I_{MSB} off.

- Cell number 2: the input current to cell 2 is still $I_{in2} = I_{in3}$. Bn_2 is pulled down, which maintains Bb_2 low. The mirror's output transistor M_C unsaturates to the value $I_{B2} = 4nA$.

- Cell number 1: the input current I_{in1} of cell 1 is then $I_{in3} - I_{B2} = 1.1nA$, which is smaller than $I_{B1} = 2nA$. The internal node Bn_1 is therefore maintained high.

- Cell number 0: the input current I_{in0} of cell 0 is still $I_{in3} - I_{B2} = 1.1nA$, which is larger than $I_{B0} = 1nA$. The internal node Bn_0 falls slowly, because of the small current difference, pulling Bb_0 low. The mirror output transistor M_C unsaturates to the value $I_{B2} + I_{B0} = 5nA$, which represents the desired quantized stored value that will be output to the synapse when mirror $M_B - M_C$ is switched to the restore phase.

9.9 Discussion

A simple long-term storage cell has been proposed which can be used to memorise the synaptic weights of analog neural networks after the learning process. A brief comparison with two other storage techniques has highlighted the context in which the storage cell can be used, taking into account their respective advantages and drawbacks. It has been shown that the cell is best suited to the implementation of neural networks with application in programmable autonomous systems, particularly if they are powered with a battery. The size of the storage cell is reasonable and it can therefore be used in practical applications where reliable long-term storage is required.

CHAPTER 10

Switched capacitor techniques

Co-author: **Stephen Pickard**

In Chapter 4 we discussed the overall decisions that must be made in designing a practical VLSI neural network. Technology, area, power, speed, memory, I/O, packaging, noise and testability all need to be considered. Two of the implementations presented there (Section 4.2.3, Section 4.2.7 and Section 4.2.8) are analysed in this case study.

The first implements all elements of a feedforward network and is based on charge manipulation. The second circuit is a variable gain neuron where control of the differential charge on a pair of capacitors results in current to voltage conversion. The operation of this neuron in a (10,6,4) network is discussed. Low voltage, single supply rail and low power operation were target characteristics for both designs.

10.1 A charge-based network

This technique was investigated because of its potential for close matching, very low power consumption, convenient weight storage and gain control with overall simplicity. A 1,1,1 net was designed as space was limited due to several test circuits sharing the same die. This was sufficient for validation of the design principles and the circuit was fabricated using a 1.2 μm, double polysilicon, double metal process. The block diagram in Figure 10.1 illustrates how the circuit functions.

We successfully tested the circuit using a 3 Volt supply with the charge amplifiers biased in the weak inversion region. The switching elements were driven by a non-overlapping, 3 phase, 400Hz clock. Faster operation is possible if bias currents are increased. Briefly, the circuit operates as follows:

- Weight and gain values are written to the synapse and neuron latches.

- During phase 1 of the clock an input signal is connected to the synaptic capacitor matrix for a period determined by the clock frequency.

- Phase 2 of the clock discharges the neuron capacitor and precharges the lower plate of the synaptic capacitor to the neuron input.

- The synaptic capacitor is then connected to a neuron through a crossover switch during the third phase.

- The neuron is then connected to the next layer, the input reconnected to the first layer, and so on, as the process is repeated.

The essential functions of a multilayer feedforward network are multiplication, addition and a non-linear transfer function. To add signals it is common to sum currents at a node. An alternative, used here, is to sum charge. This has the advantage that charge is the product of voltage and capacitance which makes design of the multiplication element straight forward. We will consider each of the functional elements in turn.

10.1.1 Synapse

The synapse used is the Charge Based Synapse described in Section 4.2.7. A weight value is represented as a variable capacitance and the input signal as a voltage. Four quadrant operation is possible because the polarity of the capacitor connection is controlled. The synapse was designed for 7 bit operation as a reasonable compromise between size and performance. The variable capacitor was realised as a matrix of unit capacitors (approximately 0.15pF) using the two polysilicon layers available. Higher capacitance per unit area may be obtained by using the gate of a FET but this is dependent upon gate voltage and cannot provide a floating capacitor. An 8×8 matrix is sufficient for 6 bit resolution and the sign bit determines the polarity of its connection to the summing node. The configuration, and thus the capacitance, was set by writing a 7 bit word into a latch.

Another important aspect of the synapse design was to maximise the dynamic range to maintain a high signal to noise ratio. This was achieved by biasing one side of the synaptic capacitor to a

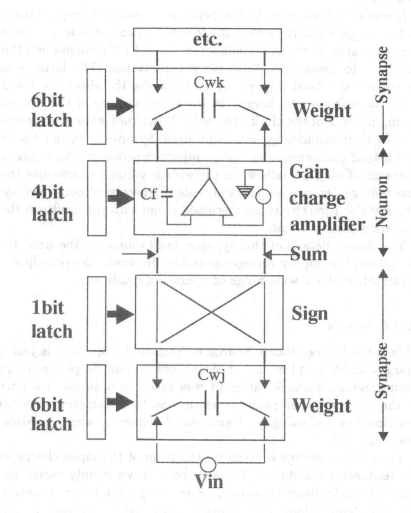

Figure 10.1 *Functionality of a charge based network including weight and gain storage. The latches configure the capacitor matrices.*

voltage approximately midway between the rails. See Figure 4.9 in Chapter 4.

It was very important to compensate for parasitic capacitances as the weight capacitors were made small to conserve area. These parasitics arise from the metallisation, the n-FET switches and the substrate to lower polysilicon plate capacitance. The latter was considered the most important. To minimise the effect the lower plate (node A) is pre-charged to the input voltage of the neuron during phase 2 of the clock. The n-FET switches were made small so that their capacitance was minimised. Symmetry in the design also helped compensate for charge injection problems. As the conductance of an FET falls when the source voltage approaches the gate voltage, the pass switches provide the necessary non-linearity. The metallisation effects are discussed along with the results in the operation section below.

The design details of this synapse are tailored to the need for low power, low supply voltage operation. However, the principle is applicable under a wide range of operating conditions.

10.1.2 Neuron

This is the Charge Based Neuron of Section 4.2.8, which is just a charge amplifier and we used a simple operational amplifier design from previous work. No attempt was made to optimise this part of the circuit as our prime concern was to demonstrate overall functionality and using tried and tested elements, where possible, minimises risk.

The output voltage is given by the ratio of the input charge to the feedback capacitance. To vary the gain we simply varied the value of the feedback capacitor, again using a latch and a matrix. For this element four bit resolution was considered adequate and allowed us to butt the synapse and neuron together. This is an important consideration in the layout of neural networks.

After each forward pass it is necessary to discharge the gain capacitor which also reduces the problem of leakage current integration. We did this by placing an n-FET in parallel that was driven by phase 2 of the clock. One has to be careful here as it is possible for the charge to be shared with the FET as it is turned on and injected back onto the gain capacitor when it is turned off. Simulations show that a small FET works best.

As we wanted four quadrant operation the output of the am-

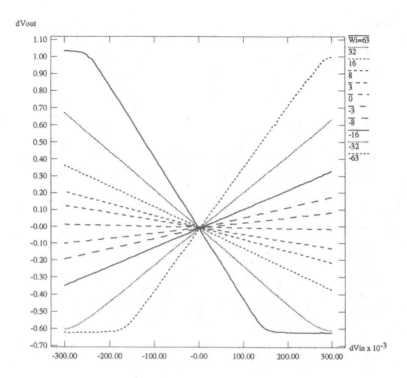

Figure 10.2 *Differential neuron output voltage (V) as a function of differential input voltage (V) for a range of weight values.*

plifier needed to have a positive and negative swing. To do this a reference voltage was applied to the positive input. See Figure 4.10 in Chapter 4.

10.1.3 Operation

The following tests were done:

- The differential input voltage was swept over ±300mV and the output recorded for input weight values ranging from -63 to +63. See Figure 10.2.

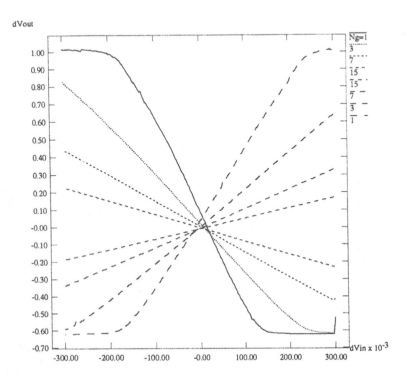

Figure 10.3 *Differential neuron output voltage (V) as a function of differential input voltage (V) for a range of weight values.*

- The differential input voltage was swept over ±300mV and the output recorded for neuron gains ranging from +1 to +15, with input weight fixed at -7 and +7. See Figure 10.3.

- The input weight was swept over the range ±63 and the output recorded for input voltages of ±20mV, ±100mV and ±300mV. See Figure 10.4.

- The above experiment was repeated for different chips. See Figure 10.5.

The results were as expected from simulations. The asymmetry in *dVout* shown in Figure 10.2 varies with the reference voltage.

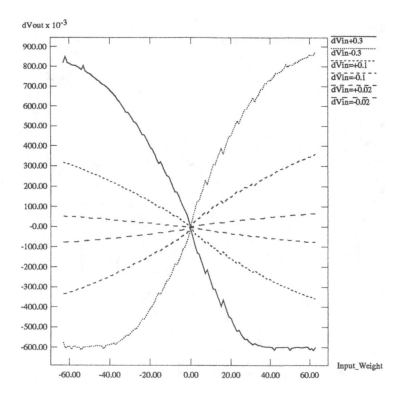

Figure 10.4 *Differential neuron output voltage (V) as a function of weight value for a range of input voltages.*

Note the large and very linear outputs, with the desired saturation occurring at the extremes. There is a slight offset of about 10mV but this will be due, in part, to the output buffers necessary for off-chip measurements. Similar results are shown in Figure 10.3 for the variation in neuron gain.

The most interesting result is shown in Figure 10.4 where close examination reveals a jagged response at high signal values. This proved to be repeatable across chips and was traced to the influence of parasitic capacitances associated with the metallisation. This

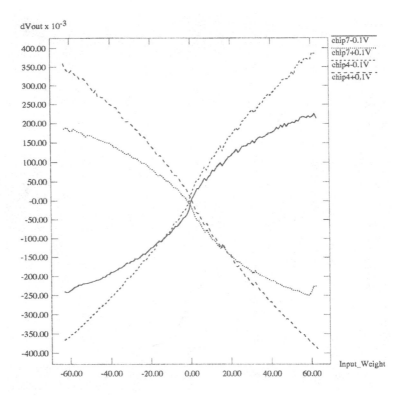

Figure 10.5 *Differential neuron output voltage (V) as a function of weight value for different die.*

lack of monotonicity could interfere with some training algorithms and so care must be taken in design to eliminate it.

10.1.4 Discussion

The circuit performed very well and met all expectations. A number of points need to be addressed though to improve future designs. The most important is to remove the metallization parasitics. A new layout of the weight matrix was simulated that moved away from the usual symmetric design. This gave us a smaller overall synapse area and included estimates of the contribution to each capacitance from the metallisation. The metallisation capacitance was also reduced as part of this redesign. It is possible to layout the synapse in an area of less than $0.04mm^2$. The neuron should be less than $0.03mm^2$ though its size is less critical.

The other element to focus on is the charge amplifier. The design of this should be optimised to take account of the operational parameters: supply voltage; dynamic range; clock frequency; and power consumption. There are many articles that cover such designs and the standard literature can be referred to [Gregorian and Temes, (1986)].

In summary the advantages of charge based design techniques include linearity, wide dynamic range, noise immunity, very low power, improved temperature tolerance, good component matching, simplicity, small area and non-volatile, on-chip weight and gain storage. The disadvantages include slow speed, relatively complex switching, difficult simulation, and limited resolution.

Device level simulation of even a small neural network is a significant problem (Sec. 5.4). Simulating switched capacitor circuits compounds this problem and so it is advisable to design circuits as conservatively as possible.

10.2 Variable gain, linear, switched capacitor neurons

In this section we focus on the design of the Switched Capacitor, Trainable Gain Neuron described in Section 4.2.3. Most analog integrated circuit implementations of the synapse are based around the Gilbert multiplier (Section 4.2.1). In the circuits used in this study the output of the multiplier is a differential current. This requires that the neuron be able to sum the input currents and provide a differential voltage output. In order to realise very low power performance, resistances of the order of 10MΩ are needed. Summing currents is trivial, one merely connects the signals at a node. Fabricating a 10MΩ resistor is not. For example, diffusion resistors would take up too much area. Table 10.1 summarises the approximate resistor values that can be realised using standard processes, within an area of $100 \times 100\mu m^2$.

If one also wishes to control neuron gain the choices are further limited. The switched capacitor technique would seem to be an obvious and straightforward choice but as we will show, there are some important design considerations.

Table 10.1 *Standard processes resistor values*

Technology	Resistance	Comments
Poly resistor	25kΩ	
Diffusion resistor	38kΩ	
n-well resistor	850kΩ	
Biased FET	2MΩ	subthreshold bias
Biased diode	10MΩ	biased @100nA
Switched capacitor	6MΩ/clock cycle	5pF @ 32kHz

10.2.1 Design considerations

For a given current, the voltage developed across a capacitor is linearly proportional to the charging time. This provides a convenient means of gain control as one merely has to vary the charging time. The circuit used is shown in Figure 10.6.

The series FET is used to control charging and the shunt FET is used to reset the capacitor after each feedforward pass. This circuit results in a linear current to voltage conversion and the required non-linearity is provided by the multiplier which has an inherent hyperbolic tangent transfer function, as described in Section 4.2.1. Another benefit is the fact that the neuron output is referenced to the supply rail. This maximises the dynamic range as the input to the next multiplier must be above about 1 Volt.

To choose the capacitor size one must optimise the area, matching requirements and charging time. An advantage of the double polysilicon process is that matched capacitors can readily be made. The uncertainty in absolute capacitance leads to an uncertainty in gain which is easily adjusted for by varying the clock duty cycle. A value of 5pF proved the most suitable for our purposes. The dimensions of the series and shunt FETs were chosen to minimise charge injection effects.

10.2.2 Operation

The switched capacitor neuron design was incorporated into a 10 : 6 : 4 two layer perceptron. The chip, called Wattle, was fabricated using a 1.2μm CMOS process. The floor plan in Figure 10.7 illustrates the chip layout.

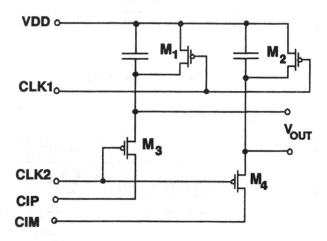

Figure 10.6 *Differential, switched capacitor neuron showing charge and reset transistors.*

The row and column address registers are used to write the weight values into the MDACs, (see Section 4.2.1). The multiplexers are used to provide the clock signals to the neurons. Two clocks are needed by the neuron: a charging clock and a reset clock. The duty cycle of the charging clock determines the neuron gain. Differential signals are used in all circuits to improve rejection of common mode noise. This also facilitates four quadrant operation from a single (3V) supply rail. Buffers are used to drive the neuron outputs off chip to avoid the effects of stray pad capacitance. The characteristics of the chip are summarised in Table 10.2.

From the circuit diagram shown earlier one might assume that the two non-overlapping clocks would be sufficient to operate the neuron. Detailed tests showed that the clocks had to be modified however, to compensate for transient effects arising in the synapses. A single current source, shown in Figure 10.7, has to provide the

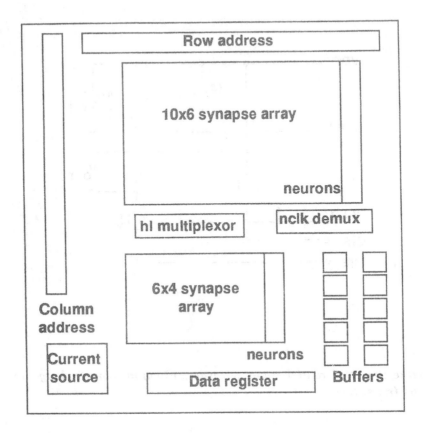

Figure 10.7 *Wattle floorplan. A 10,6,4 neural network with weighted current MDAC synapses and switched capacitor neurons.*

bias currents for all 84 synapses. A separate bias current is generated for each bit, e.g., $1nA$, $2nA$, $4nA$, etc. Each (bit) current has to charge parasitic capacitances, including gate capacitances, of the bias FETs in each synapse. The current bias generator circuit is shown in Figure 10.8 and was previously described in Section 4.2.2.

This means that after the weights have been updated as part of the training process, a short time must be allowed for the bias currents to settle. The amount of capacitance seen by the current source is a function of the total number of gates that are turned

Table 10.2 *Wattle chip characteristics*

Parameter	Value	Comments
Area	2.2x2.2 mm^2	
Technology	1.2 μm n-well CMOS 2M2P	standard process
Weight resolution	Six bit	weights on-chip
Gain resolution	Seven bit	off-chip control
Energy per connection	43pJ	all weights at max.
LSB DAC current	200pA	typical
Feedforward delay	1.5mS	at 200pA, 3V supply
Synapse offset	5mV	typical max.
Gain cross-talk	10%	maximum

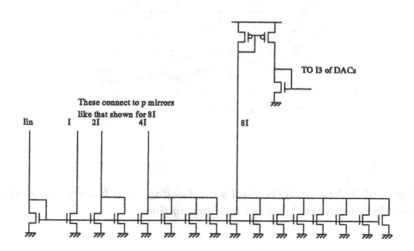

Figure 10.8 *Portion of current source (16I not shown) that supplies all MDAC synapses.* © *1995 IEEE, reproduced with permission.*

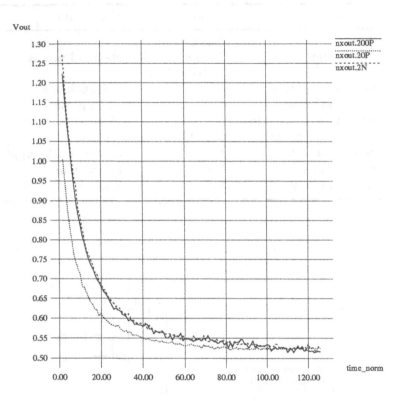

Figure 10.9 *Neuron differential output voltage as a function of overlap time (normalised with respect to charging time).*

on in the synapses which is in turn a function of the weight values. When a transistor moves from cut off to saturation the gate capacitance decreases. By overlapping the two neuron clocks the synapses are effectively temporarily connected to the supply rail during the settling time, thus incorrect charging is avoided. The overlap time is plotted in Figure 10.9 for three different bias currents.

A consequence of individual neuron gains is gain crosstalk . If neurons in the same layer have different gains then their series FETs are switched off at different times. This means that the FETs in those synapse MDACs move from saturation to cut off. As ex-

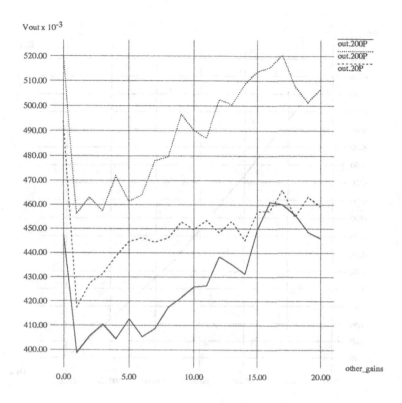

Figure 10.10 *Measurements of gain crosstalk. The output voltage of one neuron is plotted as a function of the gain on the other neurons for a range of bias currents.*

plained previously this changes the capacitive load seen by the current source and so effects the currents in the neurons that are still on. This effect was measured by fixing the gain of one hidden layer neuron at 20 and varying the others. The results are plotted in Figure 10.10 for three different bias currents. We can see that there is a worst case effect on the output of about 10%.

Neuron gain linearity (hidden layer) is shown in Figure 10.11. As can be seen it is very linear up to high gains where the common mode voltage has fallen to 1.3V causing common mode saturation

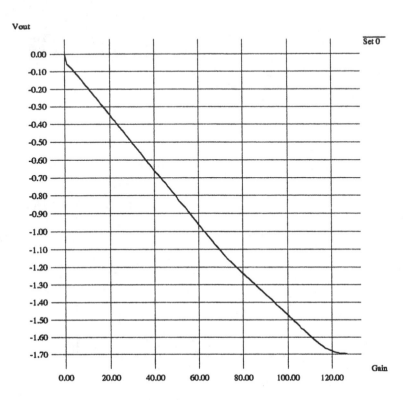

Figure 10.11 *Neuron output voltage as a function of gain showing gain linearity.*

of the following synapse. However, if one measures a single channel then a non-zero slope is measured in one quadrant. This is caused by a stray capacitance of about 0.5pF that connects between the positive and negative signals. But no detrimental effects on training could be attributed to this phenomenon.

10.2.3 Training and generalisation performance

The analog neural network using the switched capacitor based neurons , that is Wattle, was tested on a variety of problems. The results for two of these (the parity 4 problem and an ICEG training problem) are reported in this section.

The parity 4 problem converged 8 out of 10 times training the gains within a maximum of 300 iterations of CSA (see Section 6.5.5). Leaving the gains fixed, convergence was obtained 3 out of 6 times within 500 iterations. On the numerical model, parity 4 did not converge within 300 iterations. This suggests that the influence of noise may have assisted the training as in annealing based algorithms. In all cases a 5:5:1 network was used where the 5th input is a bias.

The ICEG training problem consisted of separating the ICEG morphology of patients with Retrograde Ventricular Tachycardia. The morphologies to be separated were Normal Sinus Rhythm and Ventricular Tachycardia on an individual patient basis. In this case, 10 morphology samples are used as inputs to a $10:H:2$ network where H was varied from 1 to 6. The 8 training patterns were selected and the network of each architecture was trained twice.

Table 10.3 summarises the classification performance on data not previously seen by the network for simulated and in loop chip training. Figure 10.12 shows an example of the CSA algorithm training the chip in-loop. The CSA algorithm should always reduce the error, however the effect of the noise in the hardware is evident once the training reaches smaller error values.

10.2.4 Discussion

The ability to continuously vary the gain of the neurons allowed the circuit to operate with a very wide range of bias currents: 1pA to 10nA, four orders of magnitude! Certain characteristics of the multipliers were in fact measured at bias currents below 0.1pA. Though this neuron design worked well it nevertheless requires fairly complex switching circuitry. In particular the design needs to be modified to eliminate the problems caused by gain crosstalk. Of course, this is no problem if using the same gain for each neuron in a layer is sufficient.

The need to overlap the clocks may be eliminated if one were to redesign the current source so that it is insensitive to changes in capacitive load. A simple remedy for a small network would be to place large capacitors in parallel with the gate capacitances. Another possibility is to switch dummy FETs in opposition to the ones in the MDACs so that the capacitive load remains constant.

Table 10.3 *ICEG generalisation performance*

No. of hidden units	Testing patterns correct %			
	Sim.		Chip.	
	NSR	VT	NSR	VT
1	24	100	91	99
	24	100	90	98
2	94	99	97	93
	94	99	96	93
3	97	87	100	84
	97	87	100	82
4	96	97	90	97
	96	97	91	99
5	100	99	100	97
	100	99	97	99
6	94	93	94	95
	94	93	96	97

The choice is a matter of balancing clock complexity, area and DAC complexity.

Other more minor considerations are: the input capacitance of the buffers as well as metallisation parasitics must be allowed for when selecting the neuron capacitors; layout must be as symmetrical as possible to improve the matching of the neuron capacitances; the clock speeds must be selected to allow for the minimum sized capacitors.

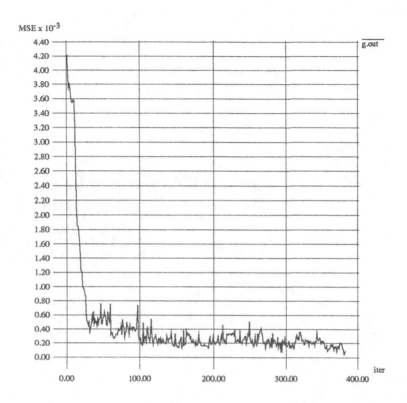

Figure 10.12 *An example training run of the analog neural network using the switched capacitor based neuron.*

CHAPTER 11

A high speed image understanding system

Eric Cosatto & Hans Peter Graf

11.1 Introduction

Vision systems, including the human eye, typically start with feature extraction to analyze an image. This is the process of taking a raw image and measuring basic characteristics, for example edges of various orientations, strokes of various thicknesses, end-lines, etc., in order to describe its content with an 'alphabet' of basic shapes. This provides a compact representation of the image content that is well suited for interpretation. Objects can be identified from the presence of a few shapes and therefore, from such a representation, a layout analysis of a scene can be done efficiently.

Extracting these basic shapes from the image is best done using convolutions, followed by thresholding. The convolution compares a part of the image with a prototype (or kernel) and thresholding decides whether there is a match. Every pixel in the image is analyzed this way and the result, called a feature map, describes the presence of the shape in the image. This method is robust against noise, in particular if the convolution kernels are large the corresponding feature maps will be free of noise. Unfortunately, this technique is also extremely computationally intensive when multiple large kernels are used. The complexity of the algorithm scales with $N \times kx \times ky \times ix \times iy$, where N is the number of kernels, kx, ky the size of the convolution kernel and ix, iy the size of the image. Special purpose hardware is needed to realize this operation in a reasonable amount of time.

We are using feature extraction in a wide variety of image analysis applications: address block location on mail pieces [Graf and Cosatto (1994)], courtesy and legal amount location on bank checks, image cleaning, forms identification and page layout analysis. In all these applications, large, complex images have to be analyzed. Feature extraction by filtering out noise and characterizing each pixel as being part of a certain shape, provides a representation from which high level analysis can be done very efficiently.

The NET32K system is an accelerator board developed for feature extraction. Two analog neural-network chips and on-board formatting/encoding capabilities are the key parts of the system. The NET32K chips store kernels and perform convolution, followed by thresholding, simultaneously with all the kernels. The board interfaces the chips with the host computer and performs formatting and encoding/decoding of the image data. The NET32K system can process up to 20 frames per second, achieving over 100 billion connections per second. Section 11.2 describes the internals of the NET32K chip and section 11.3 describes the board. Section 11.4 gives an overview of several applications where extracting multiple features with the NET32K system is the key factor for their success.

11.2 The NET32K chip

The NET32K neural-net chip [Graf and Henderson (1990)] implements 256 neurons, each with 128 synapses. Summation of the synapses' contribution is done in analog. In this way full parallelism is achieved: all neurons as well as all synapses are computed simultaneously. The NET32K chip is fabricated in $0.9\,\mu$m CMOS technology and contains 420,000 transistors on a $4.5 \times 7\,\text{mm}^2$ die. Analog computation reduces power consumption, size and cost of the chip, making it possible to implement 32K connections. The digital I/O allow the NET32K to be integrated easily in a digital system, hiding the analog elements from the user.

11.2.1 Architecture

The NET32K chip consists of three blocks: input section, connections, and output section (Fig. 11.1). The input formatter contains 16 serial-to-parallel shift registers, each with 16 bits to optimize access to input data. The chip's connections can be split in two parts: the connection matrix where the kernels are stored and the reference block where individual references for each neuron's thresholding function are stored. The output section contains the comparators and output registers.

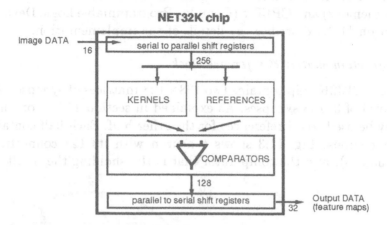

Figure 11.1 *Block Schematic of the NET32K chip.*

Input section

The input section of the NET32K (Fig. 11.2) transforms a 16 bit input into a 256 bit output using serial to parallel shift registers. This supports sliding of a 16 × 16 window over an image. To save area, these shift registers are implemented as linked dynamic memory cells. Data are shifted through the cells using a two–phase–non–overlapping clock. No refresh is necessary, since during the convolution, data are constantly shifted at a sufficiently high rate.

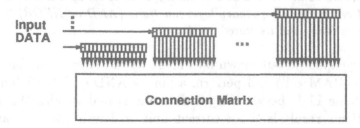

Figure 11.2 *The input section of the NET32K chip consists of 16 serial to parallel shift registers each with 16 bits.*

It is still possible to further reduce the input bandwidth using variable length delay lines. However, the amount of data storage required by this technique makes it questionable whether it should be implemented directly on the chip. Instead, we chose to implement it at the board level using off-the-shelf FIFO's (first in first out memory) and CPLD's (Complex Programmable Logic Device). Section 11.3.1 describes the details of the implementation.

Connection matrix & reference block

The NET32K chip contains two 128×128 matrices of synapses, for a total of 32768 synapses. As explained in section 11.2.2, one half may be used to set references for the other half. Each half contains 128 neurons. Fig. 11.3 shows a neuron with its 128 connections (synapses), and the comparator that is thresholding the result.

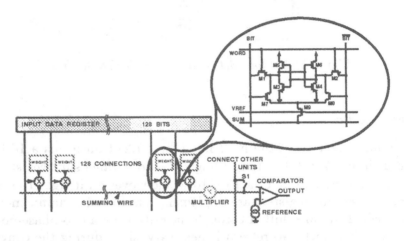

Figure 11.3 *A neuron contains 128 synapses and all their current contributions are summed on a wire in analog form. A synapse is a 1-bit static RAM cell performing binary logic functions (AND or NXOR) between the input data and its stored weight.*

A synapse (detail shown on Fig. 11.3) can store a one-bit value (static RAM cell) and perform a binary AND or NXOR function (see table 11.1) between input data and stored weight. When the function's result is '1', a current unit is driven onto the analog summation wire. The total current is transformed into a voltage that is compared with a reference voltage and thresholded. It is a binary value that indicates whether the total contribution from the connections of a neuron is bigger or smaller than the reference. All connections are implemented in the chip as SRAM cells and need no refreshing. Once new data are presented to the input of the synapses, 100 ns are necessary for the network to settle down. The thresholded response of all neurons can then be read out from the comparators. The peak computation rate of the NET32K chip is 320×10^9 connections per second.

Table 11.1 *binary NXOR and AND functions*

a	b	a NXOR b	a AND b
0	0	1	0
0	1	0	0
1	0	0	0
1	1	1	1

Output section

When the network has settled down, the results of all the comparators are latched and can be read out. Parallel to serial shift registers are used in order to read the result from the internal 128 bits wide bus through 32 output pins.

11.2.2 Configuration

The neurons' size, the resolution of the weights and the threshold values are programmable as described in the following sections. Table 11.2 lists some of the most useful configurations for the NET32K chip.

Table 11.2 *A few possible configurations of the NET32K chip.*

Threshold	Kernel size	Weight precision	# neurons to store a kernel	# kernels available
individual	16 x 16	2 levels	2	64
individual	16 x 16	3 to 4 levels	4	32
individual	16 x 16	5 to 16 levels	8	16

Neuron's size and precision

Up to 8 neurons can be grouped together to form a bigger neuron by averaging the analog current contribution of each of them. Configuration bits determine whether 2, 4 or 8 neurons are tied together (256, 512 or 1024 connections). Figure 11.4 shows how four neurons may be connected. The increased size of the neurons can be exploited to implement larger size kernels as well as to increase the precision of the weights. Fig. 11.5 shows how one can increase the precision of the weights by using two binary kernels to form a ternary kernel. Adding the contributions of the two binary kernels is equivalent to using a ternary kernel (three levels: -1, 0, 1). Four binary kernels can be used to form a kernel with weights having fifteen levels (-7, -6, ..., 0, ..., 7). Each neuron has a configurable multiplier (1/1, 1/2, 1/4 or 1/8). Therefore each contribution can be divided by 2, 4 or 8. By setting a different factor for each neuron of the same group of four, a binary encoding of the weights is obtained and 15 different levels can be coded into one kernel pixel.

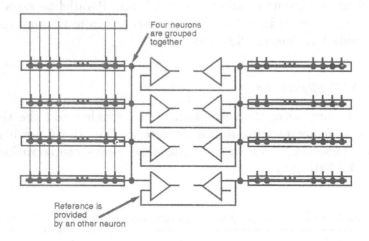

Figure 11.4 *Four neurons connected together increase the size or precision of the kernel. Connecting neurons is done at the analog level, before thresholding is applied. To obtain a precise reference value for thresholding, four neurons are connected together on the right side. By not connecting them, four different threshold references can be used with lower precision, providing a four level result.*

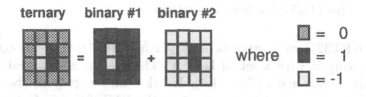

Figure 11.5 *Trading size against precision is realized by combining two binary kernels into a ternary kernel.*

Neuron's Reference

The thresholding function of a neuron is a simple step function. The threshold may be set globally (common to all neurons) or individually (each neuron has a different reference threshold). When a common threshold is selected for all neurons, the whole chip may be used to store kernels. However, if individual thresholds need to be used, half of the chip's neurons will be dedicated to store and provide the thresholds for the other half. This configuration is usually chosen since a specific response for each kernel is necessary in most applications.

11.2.3 Summary

The NET32K neural-network chip is a very powerful co-processor to achieve feature extraction on binary images. Parallel computation as well as analog techniques allow for very high performance:

- 32K connections: 16K for kernels, 16K for references.
- reconfigurable architecture.
  - 16 kernels with 1024 bits (16x16, 5 to 15 levels precision).
  - 32 kernels with 512 bits (16x16, 3 to 4 levels precision).
  - 64 kernels with 256 bits (16x16, 2 levels precision).
- I/O optimized for convolutions.
- Parallel and analog computation achieve 320 G C/s at 1 bit resolution.

The chip has been successfully integrated in several high-speed printed circuit boards and is used in various applications.

11.3 The NET32K board system

The NET32K chip extracts multiple features from an image by convolving it with a set of kernels. The system is designed as a pipelined data path (Fig. 11.6). It takes an image from the host computer, formats it and sends it to the NET32K chip to be convolved with the kernel set. The output of the chip is formatted again and then sent back to the host. A bidirectional FIFO (first in, first out), is used to ease the transfer of images from and to the host.

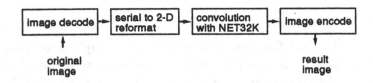

Figure 11.6 *Image data flow through the NET32K system.*

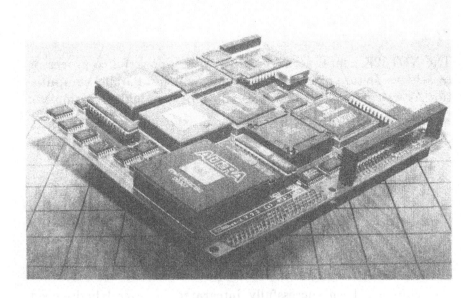

Figure 11.7 *Photograph of the NET32K SBus board.*

11.3.1 Architecture

The detailed block diagram of the NET32K board is shown in Fig. 11.8. A bidirectional FIFO serves as a buffer on the data bus. This chip contains two separate FIFO's, one for each direction of the data. Whenever data are put in its input FIFO, the board reads them, processes them according to the instruction in the control register, and writes a result into the output FIFO. Therefore, all the host has to do is to write the image to a memory location and the result can be read from there. Since the FIFO is of limited size ($2 \times 256 \times 36$ bits), one has to switch between sending data and reading results. To optimize this, a status register indicates at any time the space available in both input and output FIFO's. In this way one can send and read entire blocks of data.

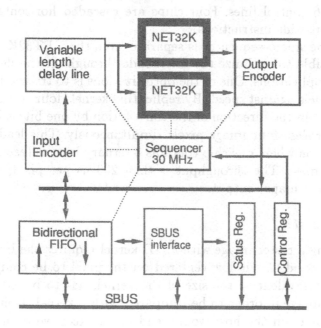

Figure 11.8 *Block level schematic of the NET32K board.*

The board is memory-mapped into the workstation address space through the SBus [SUN Microsystems (1990); Lyle (1992)]. Transfer rates of up to 20Mbytes per second are obtained easily. A microsequencer is used to control board operations and to provide instruction words to the NET32K chips. Two data formatters, one at the input and one at the output provide a flexible way to handle different data formats and compression schemes. The variable length delay line coupled with the input section of the NET32K chips transforms the one dimensional data flow into a two dimensional window sliding over the image.

Sequencer

At the heart of the system are the two NET32K chips and the sequencer. The two chips receive the same instruction, but have different weights stored in their memory. The sequencer contains the microcode that operates the NET32K chips, as well as the formatter, encoder and decoder chips. It is implemented with four ALTERA EPS448-30 clocked at 30 MHz. These chips are EPROM based micro-sequencers that are easy to interface. They are given an instruction that starts a microcode sequence. One of these chips outputs 16 control lines. Four chips are cascaded horizontally to issue 64 bit wide instructions.

Since the micro-sequencer is separate from the NET32K, it has been possible to optimize the microcode throughout the development of applications. One of the optimizations is to trade off number of kernels against speed. By replicating kernels four times, each one shifted in the direction of the convolution by one bit, it is possible to process four image pixels simultaneously. This leads to a configuration where one chip contains 8 ternary kernels (each replicated 4 times). The throughput is then 200 ns per pixel, with 8 bits (feature maps) output.

Input formatter

Convolving a raster image with a 2-D kernel requires the image to be reformatted. A window centered on the pixel to be computed, of a size equivalent to the size of the kernel, has to be extracted from the image in order to be compared with a kernel. Doing this for each pixel on the host would lead to an excessive data bandwidth between memory and the NET32K board. Since pixels are convolved one after another along a line, and one line after another over the entire image, shift registers may be used to hold a window that can slide over the image as shown in Figure 11.9.

On the board, a delay line of variable length (16 bits wide) is implemented with a FIFO and two 16-bit shift registers. Figure 11.10 shows the block diagram of the variable length delay line. This implementation is very flexible, and the length of a scanline is only limited by the depth of the FIFO. FIFO's up to 32,728 words deep

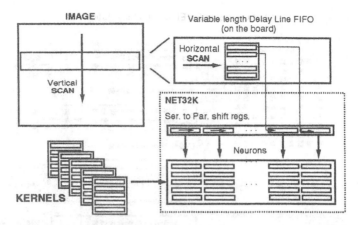

Figure 11.9 *Sliding a convolution window over an image. The variable length delay line FIFO holds a strip of the image that slides vertically over it. Within each strip, the convolution window slides horizontally. While the variable length delay line FIFO is on the board, the shift registers that hold the window are on the NET32K chip.*

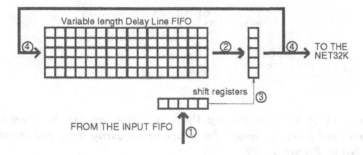

Figure 11.10 *Register level schematic of the convolution formatter. The data flow goes as follows: 1. Write a word to the shift registers. 2. Read a word from FIFO. 3. Shift one bit from input word into FIFO word. 4. write result to NET32K and back into FIFO. Repeat from step 2. until input word is empty.*

are readily available. Setting up the FIFO at the beginning of an image requires writing n words into the FIFO, where n is the length of the image. From then on, only one new pixel is loaded into the FIFO per operation done on the NET32K chips. The formatting steps are illustrated in figure 11.11.

Shifting Pixels for Efficient Convolution

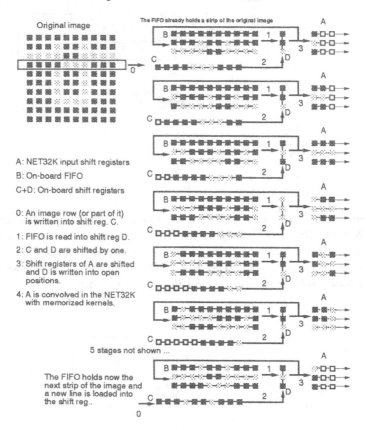

A: NET32K input shift registers

B: On-board FIFO

C+D: On-board shift registers

0: An image row (or part of it) is written into shift reg. C.

1: FIFO is read into shift reg D.

2: C and D are shifted by one.

3: Shift registers of A are shifted and D is written into open positions.

4: A is convolved in the NET32K with memorized kernels.

Figure 11.11 *A few steps through the convolution process show how pixels are extracted from an image for convolution using the pixel formatter described in Figure 11.10*

Input decoder, RLE format

The NET32K chip is optimized to convolve binary images, where one pixel of an image is represented by one bit. A common way to compress such an image is to encode black runs (connected black pixels along a row of an image) by their end-points. The runs are typically organized in a buffer with pointers to the first run of each line. The size of the encoded image can then vary between 0 (all white image) and, in the worst case: $(nx \times ny)/2$ runs, where the image alternates from black to white at each pixel. Typically an image of a check with around 1 million pixels requires a size of 20 to 50 Kbytes in RLE representation.

The board accepts these RLE words, converts them to packed bits (where each pixel is represented by one bit), and reformats the bits for the convolution. To keep track of the end-of-lines, writing to a different address on the SBus space triggers an end of line sequence that writes an end-of-line word to the output FIFO. The board also accepts images in packed bit format.

Output encoder

Since the NET32K chip convolves an image simultaneously with many kernels, each pixel on the original image produces many bits in the result image. To avoid I/O bottlenecks between board and host, one has to represent the result in a compact way. We encode several feature maps into one or more result images. The board includes a 5000 gates programmable logic device (EPLD EPM7256 from ALTERA) that is used to combine and encode the feature maps. For most of our applications, we have been using logic combinations (OR / AND) of different feature maps that are then coded in RLE format. However this is not the only possible operation. This EPLD can be reconfigured easily to serve the needs of different applications.

11.3.2 Summary

The NET32K board has been designed as a simple data-path. I/O has been carefully designed to avoid bottlenecks and allow the NET32K chips to run close to their peak speed. The following items are key to the system's high performance.

- 2 NET32K chips for over 100 billion connections per second.

- A variable length delay line (width=16, length up to FIFO depth).

- Data encoder/decoder accepting RLE (run-length-encoded) format.

- Simple slave-only SBus interface memory mapped into the host memory.

11.4 Applications

We are applying our convolution technique with the NET32K board to several document analysis tasks and we present here three different applications. The basic sequence of operation is:

- The original image is filtered by convolving it with kernels. These kernels are designed to extract simple geometric shapes from the image such as edges, lines, corners, text lines, etc. This computationally intensive task is achieved efficiently on the NET32K hardware.

- The results are feature maps where the pixels mark the presence of features of interest. Connected component analysis is used to translate these feature maps into symbolic format.

- The feature representation of the image is used to find objects of interest or to determine the layout of a text page. These high level algorithms are typically not computationally intensive and are done on the host computer.

The connected component technique determines touching pixels in an image and labels them as being part of the same object. The required computation is data dependent. If the image is noisy and thus contains many small connected objects, the analysis may be time consuming. On the other hand, if the image has been filtered properly to only retain desired features, there are fewer connected components left and the analysis can be executed on a general purpose processor in a short time.

In the following application, **ternary** kernels are used for the convolutions. Each pixel of a ternary kernel can take three values: +1, -1, 0 ('don't care' pixel), and is internally represented by two bits: +1 by (1,1), -1 by (0,0) and 0 by (1,0) or (0,1). During the convolution, the same data bit is presented to both bits of a kernel pixel. The function performed between kernel bits and image bits is NXOR and the truth table is shown in Table 11.3.

11.4.1 Discrimination of machine printed and handwritten text

In many applications such as address readers or check processing systems, fast and robust discrimination between machine printed and handwritten text is needed.

For this purpose, we use two sets of kernels. The first set is designed to pick up strokes of different orientations (Fig. 11.13).

Table 11.3 *Truth table for convolution with ternary kernels.*

image bit	kernel (bit 1)	kernel (bit 2)	result contribution (in current units)
0	0	0	2 (match)
0	0	1	1 (don't care)
0	1	0	1 (don't care)
0	1	1	0 (no match)
1	0	0	0 (no match)
1	0	1	1 (don't care)
1	1	0	1 (don't care)
1	1	1	2 (match)

Figure 11.12 *The original image contains machine printed text as well as handwritten text. The NET32K system discriminates between the two in just one pass of convolution.*

These strokes are the building elements of handwritten text. When the image is convolved with these kernels, noise, texture and machine printed text are mostly filtered out while the positions of strokes are being marked.

The second set of kernels (Fig. 11.14) are designed to detect machine printed text. They are actually detecting any thick black line on a white background. By using a low threshold for the corresponding neurons, we are able to tune the kernel's response so that they will turn ON even though the pixels on the image and those of the kernels match only partially. Despite only a partial match, we can still discriminate well between strokes or noise and machine-printed text lines.

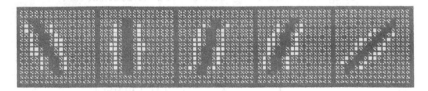

Figure 11.13 *Kernels used for detecting handwriting style. Strokes of different orientation mark the presence of handwritten text. Each pixel of a kernel can take one of three values: +1 (black), -1 (white), 0 (grey) and is internally represented by two bits.*

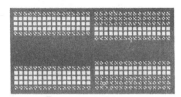

Figure 11.14 *Kernels used to detect machine printed text lines.*

The results are feature maps for each set of kernels that show where the kernels have responded strongly. This indicates the presence of a certain shape in a particular position (Fig. 11.15 and Fig. 11.16). The feature maps are processed with a connected component analysis, filtered and smeared horizontally. A clustering process provides the position of individual lines or groups of lines. In this way, we are able to locate blocks of text and label them as handwritten or machine printed.

We have been using our system with success for the address block location task [Graf and Cosatto (1994)] of an automatic address reader that has been tested by the U.S. Postal Service. The system has been trained with a large database of over 5000 mail-pieces. It is able to discriminate between machine-printed and handwritten addresses with an accuracy of 95%.

11.4.2 Noise removal

Images of mail-pieces as well as those of checks often have random noise or background texture around the text of interest (Fig. 11.17).

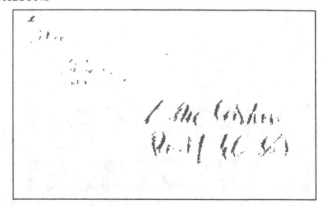

Figure 11.15 *Combination (logic-OR) of the stroke feature maps. The handwritten text responds strongly, while the machine-printed text produces only a weak response.*

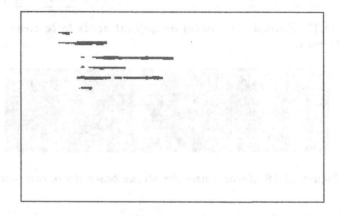

Figure 11.16 *Combination (logic-OR) of the printed-text feature maps. Here, machine-printed text lines respond well, while handwritten strokes are filtered out.*

This can be due to dirt on the paper, or, what is more often the case, noise is introduced when the image is thresholded (converted to binary). This noise prevents the optical character recognizer from reading the text, and thus the images have to be cleaned before being sent to the reader.

The kernels used here (Fig. 11.18) are designed to pick up the strokes of machine-printed characters. All the feature maps are then OR-ed together to rebuild the character (Fig. 11.19). The shape, orientation and thickness of the strokes are used to filter

Figure 11.17 *Example of a noisy image that needs to be cleaned before it can be read.*

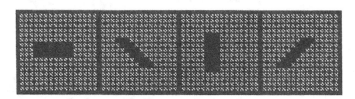

Figure 11.18 *Kernels used for stroke based noise removal.*

out the noise, which is smaller and different in shape compared to the characters. This technique is also good for filtering out non-random noise such as textures or graphical elements, for example lines, because not only the size of the noise is analyzed but also its shape. Everything that does not look like a stroke is dismissed. The kernels used here do not have white pixels, which means that all characters with strokes thicker than the kernels will be picked up.

11.4.3 Line removal

Text readers are usually sensitive to the presence of graphical lines drawn across a page. Removing those lines is desirable before starting any optical character recognition (OCR) on the image. Fig-

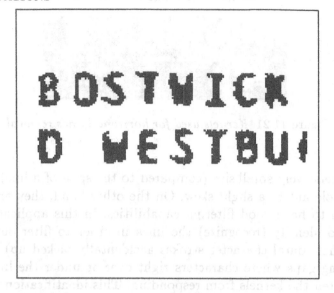

Figure 11.19 *The feature maps are combined to reconstruct the characters. The shapes of the strokes filter out noise.*

ure 11.20 shows the image with text and horizontal lines separating
the text lines. The convolution approach is well suited for removing
these lines.

Direct	Day	2	.47
Direct	Day	2	.53
Direct	Day	3	.57
Direct	Day	2	.65
Direct	Day	3	.57
Direct	Day	2	.35
Direct	Day	4	.73
Direct	Day	3	.66
Direct	Day	2	.41
Direct	Day	3	.57
Direct	Day	4	.97

continues

Figure 11.20 *The original image is a part of a telephone bill. The horizontal lines have to be removed.*

The kernels shown on Fig. 11.21 will pick up the horizontal lines
while most of the text is ignored (only long horizontal strokes are
also picked up). Different thicknesses are coded into the kernels.

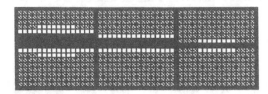

Figure 11.21 *Kernels used for horizontal lines removal.*

Their relatively small size (compared to the span of a line) makes them tolerant to a slight skew. On the other hand, they are large enough to have good filtering capabilities. In this application we need to identify (recognize) the lines in order to filter out short ones (horizontal character strokes accidentally picked up) as well as filling gaps where characters right over or under the line have prevented the kernels from responding. This identification is done by applying the connected component analysis to label the blobs in the feature map. Since the feature maps coming from the board are already in RLE format (see section 11.3.1) this is done very fast on the host processor. The lines are finally removed by overlaying white lines over the black lines in the original image (Fig. 11.23).

Figure 11.22 *Feature map after convolution with line detectors. Characters are filtered out and connected component analysis can be applied to identify long lines.*

The typical size of an image is 1700 × 2000 binary pixels (3.4 Million bits). The system performs line removal at a speed of 1 image per second, which includes convolution, connected component

Direct	Day	2	.47
Direct	Day	2	.53
Direct	Day	3	.57
Direct	Day	2	.65
Direct	Day	3	.57
Direct	Day	2	.35
Direct	Day	4	.73
Direct	Day	3	.66
Direct	Day	2	.41
Direct	Day	3	.57
Direct	Day	4	.97

continues

Figure 11.23 *After having identified the lines from the feature map, they are removed by overlaying white pixels on top of them.*

analysis, line extraction and line removal. This is about 30 times faster than a previous all-software approach.

11.5 Summary and conclusions

As challenging as their conception, is the integration of neural net chips into a system suited for an industrial environment. The parallel architecture of these chips has to be supported by a high data I/O bandwidth throughout the system, to fully exploit their computation bandwidth. Keeping cost and complexity of such a system at a reasonable level is difficult. In fact, the limited success of neural net hardware so far, is due mainly to the high cost and complexity of the available systems.

Another problem lies in the degree of specialization of such a system. Fully implemented neural networks such as the Synaptics I-1000 check reader chip [Hammerstrom (1993)] are efficient and low in cost, but not flexible. Since they are developed for one particular task, their life cycle is limited. Trying to keep a neural network system too general, on the other hand, reduces its compute speed, and it may therefore not be much faster than a powerful digital signal processor (DSP) or reduced instruction set processor (RISC). If special purpose hardware can be replaced by a general purpose processor without loosing significantly in performance, even at a higher cost, it is probably wise to avoid the inherent lack of upgradability and limited support available for a specialized piece of hardware.

Vision problems can often be divided into two distinct tasks. One is to analyze the raw pixel data to extract basic shapes that are the building blocks of more complex objects. For this task, a parallel approach using neural networks and analog computation still outperforms conventional computers by orders of magnitude. On the other hand, high level processing and interpretation of these building blocks require a more complex type of computing power (graph search, etc.) that is best handled by a general purpose processor with a powerful programming language and a compiler. The need is evident for a tightly coupled combination of special purpose hardware and conventional processors.

The NET32K SBus system provides a compact and cost effective solution for image analysis tasks such as multiple features extraction. Analog and low resolution computation inside the NET32K chip has made it possible to integrate 64 16 × 16 pixel convolvers into one small, low power chip. Due to a fast memory mapped interface with the host computer and on-board formatting and encoding/decoding capabilities, I/O bottlenecks are avoided. The NET32K system can process up to 20 frames per second (512 × 512 pixels), achieving over *100 GC/s* (billion connections per second), and is the enabling element for the image understanding algorithms described in this paper.

The key advantage of the NET32K system over other existing convolvers is the use of low resolution computation. In order to extract basic shapes from an image, reduced precision is sufficient and using bi-level images and tri-level kernels makes it possible to implement analog techniques on the chip without requiring costly analog to digital conversion (or worse, analog I/O). In the same manner, thresholding the result of the convolution on chip reduces the output bandwidth and allows multiple kernels to be used simultaneously, which is the key to solving many computer vision problems.

CHAPTER 12

A Boltzmann learning system

Joshua Alspector

12.1 Introduction

Learning in neural networks has been a bottleneck requiring huge
amounts of computation time. Furthermore, neural network models
draw their inspiration from biological nervous systems where parallel computation based on the physical characteristics of the neural
substrate is used to advantage. Motivated by these considerations,
we have built an experimental prototype learning system based on
the neural model called the Boltzmann Machine. This model has
been implemented in an analog VLSI experimental prototype and
uses the physics of electronics to advantage. It has been incorporated into a learning co-processor for standard digital computer
systems. However, it is probably best suited for embedded applications where the analog nature can be used to advantage. This
has been demonstrated in its application to adaptive non-linear
equalization for wireless communications systems. The VLSI chips
are cascadable and suitable for scaling to large systems.

12.1.1 Neural style and digital computation

Most current information processing techniques rely on computer-
like handling of digital information. The nervous system, however,
is not a state machine, nor does it include an arithmetic logic unit
(ALU) module for arithmetic operations. It is, instead, a physical
dynamical system whose state evolves in time in a complex way
guided by the physics of the neural substrate. This style of processing incorporates a physical computation whereby each neuron
evaluates and ouputs to many other neurons a non-linear function
of the sum of its inputs. Electronically, this means taking advan-

tage of charge conservation and device physics to perform such a computation.

While one can perform such a computation digitally to emulate neural processing, this does not take advantage of the physics. Our point of view is that if it's neural, it's physical and therefore implementable in physical media. Also, while many forms of analog computation take direct advantage of the physics of devices, they may not have the dynamics which would qualify as emulating the nervous system.

Neural networks are useful for handling complex tasks of various types which are difficult to solve using traditional techniques. There is the complexity of a scene image composed of a million pixels that have to be ordered and transmitted many times a second. Neural processing brings order and data reduction to this cacophony of sensory input. There is also the complexity of representation. The same physical object can be represented in many different ways as pixel patterns but an efficient representation classifies them in a common way with perhaps different orientations or shadings. There may also be the complexity of dealing with images for which there is not yet a classification or efficient representation. With enough statistical analysis or 'experience', such a representation will emerge. As we will see, the essence of handling complexity in the neural style is using correlation in stimulae to process information, represent data, and learn. Each processing step reduces the informational entropy to be presented to the next stage.

There has been a resurgence of interest in connectionist models of brain function in recent years. These models will be analyzed and the essential features necessary for VLSI implementation will be extracted.

12.1.2 Historical background

The roots of current work on neural models can be found in a 1943 paper by McCulloch and Pitts [McCulloch and Pitts (1943)]. There the brain is modeled as a collection of neurons with two states: $s_i = 0$ (not firing) and $s_i = 1$ (firing at maximum rate). If there is a connection to neuron i from neuron j, the strength of this connection is defined as w_{ij}. Each neuron readjusts its state asynchronously according to the *threshold rule*:

$$s_i = \begin{bmatrix} 1 \\ 0 \end{bmatrix} \qquad if \sum_i w_{ij} s_j \begin{bmatrix} > \\ < \end{bmatrix} \theta_i \qquad (12.1)$$

where θ_i is the threshold for neuron i to fire.

A model of this sort formed the basis for the perceptron [Rosenblatt (1961)] which is a parallel machine that can learn by example. The perceptron consists of an input array hard-wired to a set of feature detectors whose output can be an arbitrary function of the inputs. These are connected through a layer of connections with modifiable strengths or weights to threshold logic units, each of which decides whether a particular input pattern is present or absent using the *threshold rule* of Equation. 12.1 . The two components of this machine can be implemented in hardware. The threshold logic unit can be a bistable circuit such as a Schmitt trigger where a change of logic state occurs if a certain threshold is reached. It could also be an operational amplifier where high non-linear gain performs the threshold function. The adaptive weights can be implemented as variable resistors. A machine of this sort, the ADALINE (adaptive linear element) was made in the 1960's [Widrow and Hoff (1960)]. These two elements will also form the basis for our VLSI system as will be discussed later.

There exists an algorithm, the perceptron convergence procedure, which adjusts the adaptive weights between the feature analyzers and the decision unit. This procedure is guaranteed to find a solution to a pattern classification problem, if one exists, using only the single set of modifiable weights [Minsky and Papert (1969)]. Unfortunately, there is a large class of problems, which perceptrons cannot solve, as pointed out by Minsky and Papert (1969), namely those which have an order of predicate greater than 1. The boolean operation of exclusive-or has order 2, for example. Also, the perceptron convergence procedure does not apply to networks in which there is more than one layer of modifiable weights between inputs and outputs, because there is no way to decide which weights to change when an error is made. This is the 'credit assignment' problem and was a major stumbling block until recent progress in learning algorithms for multi-level machines.

Another seminal idea in brain models, also published in the 1940's, was Hebb's proposal for neural learning [Hebb (1949)]. This states that if one neuron repeatedly fires in concert with another, some change takes place in the connecting synapse to increase the efficiency of such firing. This correlational synapse postulate has, in various forms, become the basis for models of distributed associative memory [Anderson, Silverstein Ritz and Jones (1977); Kohonen (1977)]. Similar approaches also lead to learning of patterns [Grossberg (1969)].

Various neural transfer functions have been used in models. The all-or-none McCulloch-Pitts neuron is represented by a step at the threshold. The linear model is simply proportional to the sum of inputs. A real neuron exhibits saturation of the firing rate as well as a threshold with an approximately linear behavior in between and is often represented by a sigmoid function [Grossberg (1973); Sejnowski (1981)]. Any of these transfer functions can be well modeled by an electronic system. An operational amplifier has a transfer function which is close to the sigmoid. With proper gain control and feedback, it can be nearly a step or can be adjusted to operate in the linear region. We will take advantage of this in our electronic model.

12.1.3 Hopfield's model of associative memory

Information storage

The recent activity in neural network models was stimulated in large part by a non-linear model of associative memory due to Hopfield (1982). These neurons are an all-or-none type with a threshold assumed to be zero. Memories, consisting of binary vectors $s^{(k)} = (s_1^{(k)}, \ldots, s_N^{(k)})$, labeled k, are stored in the outer product sum over states

$$w_{ij} = \sum_k \left(2s_i^{(k)} - 1\right)\left(2s_j^{(k)} - 1\right) \qquad (12.2)$$

where the $(2s - 1)$ terms have the effect of transforming the $(0,1)$ neural states to $(-1,1)$ states. We see that for a particular memory $s^{(1)}$:

$$s^{(1)} = \sum_j w_{ij} s_j^{(1)} \qquad (12.3)$$

$$= \sum_{j \neq i} \left[\left(2s_i^{(1)} - 1\right)\left(2s_j^{(1)} - 1\right)\right] s_j$$

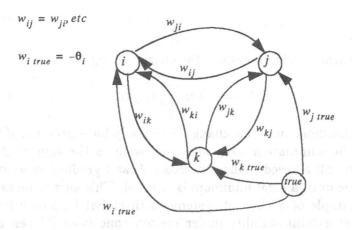

Figure 12.1 *Symmetric Hopfield network*

$$+ \sum_{k \neq l} \left(2s_i^{(k)} - 1\right)\left(2s_j^{(k)} - 1\right)s_j \right]$$

The first summation term has mean value $\dfrac{(N-1)}{2}$ for the j terms summed over N neurons while the last term in brackets has mean value zero for random (and therefore pseudo-orthogonal) memories when the sum over M memories (label k) is taken. Thus

$$\sum_j w_{ij} s_j^{(1)} \approx \frac{N-1}{2}\left(2s_i^{(1)} - 1\right) \qquad (12.4)$$

Since this is positive ($> \theta_i = 0$) if $s_i^{(1)} = 1$ and negative if $s_i^{(1)} = 0$, the state does not change under the threshold rule and is stable except for the statistical noise coming from states $k \neq 1$, which has a variance of $\sqrt[2]{\dfrac{[(M-1)(N-1)]}{2}}$.

The symmetric case and the energy measure

The proposed network (see Figure 12.1) is fully connected and symmetric. That is, for every pair of neurons labeled i and j, $w_{ij} = w_{ji}$, but $w_{ii} = 0$. Using an analogy from physics, namely the Ising model of a spin-glass [Kirkpatrick and Sherrington (1978)], we can define an energy

$$E = -\frac{1}{2} \sum_i \sum_{j \neq i} w_{ij} s_i s_j. \qquad (12.5)$$

If one neuron s_k changes state, the energy change is

$$\Delta E_k = -\Delta s_k \sum_{j \neq k} w_{kj} s_k. \qquad (12.6)$$

By the threshold rule, this change could only have occurred if the sign of the summation term were the same as the sign of Δs_k. Therefore, all allowed changes decrease E and gradient descent is automatic until a local minimum is reached. This energy measure is an example of a class of systems with global Liapunov functions that exhibit stability under certain conditions [Cohen and Grossberg (1983)]. The neural states at these minima represent the memories of the system. This is a dynamical system which, in the process of relaxation, performs a collective computation.

This model stimulated the construction of integrated circuits implementing this type of associative memory by groups at Cal Tech [Sivilotti, Emerling and Mead (1985)] and AT&T Bell Laboratories [Graf et al. (1986)]. A system of N neurons has $O(N/\log N)$ stable states and can store about 0.14 N memories ($N \sim 100$) before noise terms make it forget and make errors. Furthermore, as the system nears capacity, many spurious stable states also creep into the system, representing fraudulent memories. The search for local minima demands that the memories be uncorrelated, but correlations and generalizations therefrom are the essence of learning. A true learning machine, which is our goal, must establish these correlations by creating "internal representations" and searching for global minima; thereby solving a constraint satisfaction problem where the weights are constraints and the neural units represent features.

12.2 The Boltzmann machine

12.2.1 Internal representations

Perceptrons were limited in capability because they could only solve problems that were first order in their feature analyzers. If, however, extra layers of neurons are introduced between the input and output layers as in Figure 12.2, higher order problems such as exclusive-or can be solved by having the 'hidden units' con-

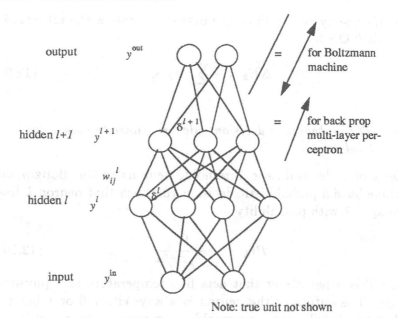

output y^{out} =   for Boltzmann
                                                                  machine

                                                             =   for back prop
hidden $l+1$ y^{l+1} δ^{l+1} multi-layer per-
                                                                  ceptron

                w_{ij}^{l}

hidden l y^{l} δ^{l}

input y^{in}

Note: true unit not shown

Figure 12.2 *A layered neural network*

struct or 'learn' internal representations appropriate for solving the problem. The Boltzmann machine [Ackley, Hinton and Sejnowski (1985)] can have this general architecture. Unlike the strictly feed-forward nature of the perceptron, connections between neurons run both ways and with equal connection strengths, as in the Hopfield model. This assures that the network can settle by gradient descent in the energy measure

$$E = -\frac{1}{2}\sum_{i}\sum_{j\neq i} w_{ij}s_i s_j + \sum_{i}\theta_i s_i \qquad (12.7)$$

where again the θ_i are the thresholds of the neurons. These threshold terms can be eliminated by assuming each neuron connects to a permanently on 'true unit' with $s_{true} = 1$ with a connection strength $w_{itrue} = -\theta_i$ to neuron i. Thus, the energy may be restated as

$$E = -\sum_{j}\sum_{i<j} w_{ij}s_i s_j \qquad (12.8)$$

while the energy gap or difference between a state with unit k OFF and unit k ON is

$$\Delta E_k = -\sum_i w_{ki} s_i \tag{12.9}$$

12.2.2 The probabilistic decision rule uses noise to escape from local minima

Instead of a deterministic threshold, neurons in the Boltzmann machine have a probabilistic decision rule such that neuron k has state $s_k = 1$ with probability

$$Pr_k = \frac{1}{1 + e^{-\frac{\Delta E_k}{T}}} \tag{12.10}$$

where T is a parameter that acts like temperature in a physical system. The output of the neuron is always either 0 or 1 but its probability distribution is a sigmoid so, on average, its output looks like the sigmoid. Note that as T approaches 0, this distribution reduces to a step function. This rule allows the system to jump occasionally to a higher energy configuration [Metropolis, Rosenbluth, Rosenbluth, Teller and Teller (1953)] and therefore to escape from local minima. The machine gets its name from the mathematical properties in thermodynamics set forth by Boltzmann and useful in optimization [Kirkpatrick, Gelatt and Vecchi (1983)].

While the Hopfield model uses local minima as the memories of the system, the Boltzmann machine uses simulated annealing to reach a global energy minimum since the relative probability of two global states α and β follows the Boltzmann distribution:

$$\frac{Pr_\alpha}{Pr_\beta} = e^{-\frac{(E_\alpha - E_\beta)}{T}} \tag{12.11}$$

and thus the lowest energy state is most probable at any temperature. The long time to thermal equilibrium at low temperatures calls for an annealing schedule that starts at a high temperature and gradually reduces it [Binder (1978)].

This is completely analogous to the physical process of annealing damage in a crystal where a high temperature causes dislocated atoms to jump around to find their lowest energy state within the crystal lattice. As the temperature is reduced the atoms lock into their proper places. The computation of such annealing on serial computers is tedious for two reasons. First, the calculation involves imposing probability distributions and physical laws on the motion of particles. Second, the calculation is serial. A physical crystal's atoms naturally obey physical laws without calculation and they all obey these laws in parallel. For the same reasons, Boltzmann machine simulations on digital computers are also tedious. Our computer simulations use Equation. 12.10 to calculate the ON probability of neurons, while we use electronic noise mechanisms to jitter the ON probability in our electronic model.

12.2.3 Electronic implementation of noise

In designing an electronic system to model real neurons, we should also take advantage of the physical behavior of electronics and use parallelism as much as possible. The sigmoidal probability distribution in Equation. 12.10 has a close electronic analog in a noisy voltage step. The probability for a neuron to be ON using the sigmoid distribution is the same within a few percent as the probability for a deterministic 'step' neuron to be on when its threshold is smeared by Gaussian noise. So another way of looking at annealing is to start with a noisy threshold and gradually reduce the noise. We have tried to take advantage of thermal noise [Motchenbacher and Fitchen (1973)] in electronic circuits which follows a Gaussian distribution for one of our chip designs [Alspector, Gupta and Allen (1989)] but later settled on a more robust method based on linear feedback shift registers [Alspector, Gannett, Haber, Parker and Chu (1991)]. The parallel generation of electronic noise can make a true Boltzmann machine in hardware. We have also incorporated the annealing effect of the noise without actually using noise by incorporating a mean-field version [Peterson and Anderson (1987)] of the Boltzmann machine in our system. These features will be described later and have been previously reported [Alspector, Jayakumar and Luna (1992)]. The enormous speedup of the electronic computation is accomplished by having the neurons evaluate their states in parallel continuously and asynchronously and by having all the connection strengths adjusted by separate processors in parallel.

12.2.4 The learning algorithm and the importance of local information.

The 'credit assignment' problem that blocked progress in multi-layer perceptrons can be solved in the Boltzmann machine framework by changing weights in such a way that only local information is used. The learning algorithm works in two phases. In phase 'plus', the input and output units are clamped to a particular pattern that is desired to be learned while the network relaxes to a state of low energy aided by an appropriately chosen annealing schedule. In phase 'minus', the output units are unclamped and the system also relaxes to a low energy state while keeping the inputs clamped. The goal of the learning algorithm is to find a set of weights such that the 'learned' outputs in the minus phase match the desired outputs in the plus phase as nearly as possible. The probability that two neurons i and j are both on in the plus phase, P_{ij}^+, can be determined by counting the number of times they are both activated averaged across some or all patterns (input-output mappings) in the training set. For each mapping, co-occurrence statistics are also collected for the minus phase to determine P_{ij}^-. Both sets of statistics are collected at thermal equilibrium. After sufficient statistics are collected, the weights are then updated according to:

$$\Delta w_{ij} = \eta \left(P_{ij}^+ - P_{ij}^- \right) \qquad (12.12)$$

where η scales the size of each weight change.

It can be shown [Ackley, Hinton and Sejnowski (1985)] that this algorithm minimizes an information theoretic measure of the discrepancy between the probabilities in the plus and minus states. It performs gradient descent in the objective function

$$G = \sum_\alpha P_\alpha^+ \log \frac{P_\alpha^+}{P_\alpha^-} \qquad (12.13)$$

where α is summed over the global states of the system. P_α^+ and P_α^- are the probabilities for those states in the clamped and unclamped phases respectively. The learning rule for general gradient descent is

$$\Delta w_{ij} = -\eta \frac{\partial G}{\partial w_{ij}} \qquad (12.14)$$

which, it can be shown [Ackley, Hinton and Sejnowski (1985)], produces the Boltzmann machine learning rule of Equation. 12.12. Minimizing G teaches the system to give the desired outputs. An important point about this procedure is that it uses only locally available information, the states of two neurons, to decide how to update the weight of the synapse connecting them. This makes possible a VLSI model where weights can be updated in parallel without any global information except whether the teacher is clamping and yet optimize a global measure of learning.

12.3 Deterministic learning by error propagation in feedforward nets

12.3.1 The generalized delta rule

The most commonly used deterministic algorithm for feedforward nets, back-propagation [Rumelhart, Hinton and Williams (1986)], takes less computer time than the Boltzmann algorithm for solving most problems. It does not require the settling to equilibrium of the Boltzmann machine at each pattern presentation. We will later see how the settling procedure can be speeded up in the mean-field version of the Boltzmann machine. The back-propagation algorithm also uses a generalization of the perceptron convergence procedure in a variation due to Widrow and Hoff (1960) called the delta rule.

This rule is applied to layered feedforward nets of the type shown in Figure 12.2 where the connections between neurons work only in the feedforward direction. The neurons have a graded semilinear response such as the sigmoid, where the output, y_i, is a differentiable function of the total net input, x_i to neuron i. This function y is just

$$y = \frac{1}{1 + e^{-x_i}} \tag{12.15}$$

for the sigmoid. For discrete (0,1) states, x_i has been called ΔE_i because in the Boltzmann machine, the energy difference is equal to the total net input. If we index each layer by the superscript l, then:

$$y_i^{l+1} = y\left(\sum_j w_{ij}^l y_j^l\right) = y\left(\Delta E_i^l\right) \tag{12.16}$$

If there are l layers of weights, there are $l + 1$ layers of units.

The perceptron rule for adjusting weights after presentation of an input-output pair is:

$$\Delta w_{ij}^{l} = \eta \delta_{i}^{l+1} y_{j}^{l} \tag{12.17}$$

where δ_{i}^{l+1} is the error in layer $l+1$, defined in Equation 12.23 and

$$\delta_{i}^{out} = \left(y_{i}^{out^{+}} - y_{i}^{out^{-}} \right) \tag{12.18}$$

is the error signal for the output layer. This is the only layer that can have an error signal in the one layer perceptron. Here y_{i}^{+} is the desired or target output at neuron i, y_{i}^{-} is the actual output, y_j is the output of neuron j giving input to i, and η scales the weight change. This rule with δ_i equal to the output error applies when there are no hidden units and the inputs feed the outputs directly.

When there are hidden units, one can obtain a learning rule that performs gradient descent on an error measure on the output units. Most often, it is the squared error objective function

$$\varepsilon = \sum_{i-outputs} \left(\delta_{i}^{out} \right)^{2} \tag{12.19}$$

that is minimized.

Adjusting output weights in the negative gradient direction gives

$$\Delta w_{ij} = -\eta \frac{\partial \varepsilon}{\partial w_{ij}}. \tag{12.20}$$

For the squared error objective function, the output error is:

$$\delta_{i}^{out} = \left(y_{i}^{out^{+}} - y_{i}^{out^{-}} \right) y_{i}^{\prime out}(x_{i}) \tag{12.21}$$

where $y_{i}^{\prime}$ is the first derivative of the semilinear activation function. If one uses an objective function more like the information theoretic measure of the Boltzmann machine,

$$G = \sum_{i-out} y_{i}^{out+} \log \frac{y_{i}^{out+}}{y_{i}^{out-}}, \tag{12.22}$$

then Equation 12.18 gives the output error. Using the appropriate δ_{i}^{out} in Equation 12.17 gives the rule for adjusting the weights feeding the output layer.

Now the error signal is propagated back to adjust the weights feeding the hidden layers. To get the error signal for the hidden units in layer l:

$$\delta_i^l = y'^l_i(x_i) \sum_k \delta_k^{l+1} w_{ki}^l \qquad (12.23)$$

where the sum is over the error signals of the units which the unit in question feeds. This applies whenever the unit is not an output unit for however many layers of hidden units may exist. Equation 12.17 is used to adjust these weights. Input units, of course, have no error.

The procedure involves first propagating the input pattern forward to compute output values y^-. The output is then compared with the targets y^+ giving the error signals δ for each output unit. Next, the error signal is recursively propagated backward, with the weights updated accordingly.

12.3.2 Analog VLSI implementation of learning

In deciding on a neural learning algorithm for analog VLSI implementation, one observes how either back-propagation or the Boltzmann machine algorithm maps into physical neurons and synapses. Such a mapping is depicted in Figure 12.3. Note that the Boltzmann neuron, panel (a), sums input (which can be implemented by current summing using the physics of charge conservation or Kirchoff's law) and uses that as input to a non-linear activation function (which can be implemented as a transimpedance amplifier). The back-propagation neuron also has this standard summing and activation, but to physically implement the algorithm, one needs an additional virtual neuron and virtual connections that implement the backward paths for error propagation as illustrated in panel (b).

This virtual neuron receives backward propagating error signals which are summed (and so can also be physically implemented with current summing). However, this summed error must now be multiplied by the derivative of the activation function. Taking an exact derivative is difficult to implement in analog VLSI.

Panel (c) shows the function of a physical synapse in the Boltzmann machine. It simply multiplies the activation of the neurons it connects to (in either direction) by the stored weight and outputs a contribution to the net input of the receiving neuron. The input of the synapse can be a voltage (which is the neuron output for easy distribution to many synapses) while the output is a current (for easy summation at the neuron input). In the Boltzmann ma-

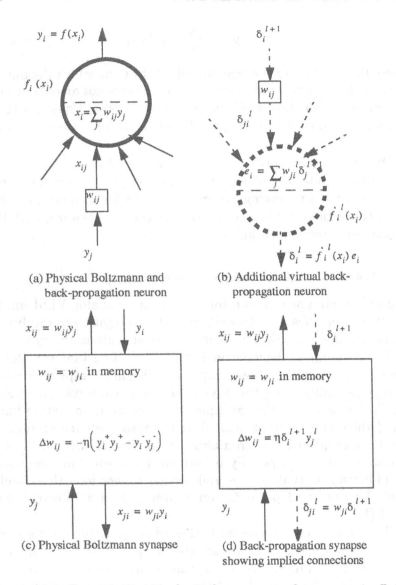

(a) Physical Boltzmann and
back-propagation neuron

(b) Additional virtual back-
propagation neuron

(c) Physical Boltzmann synapse

(d) Back-propagation synapse
showing implied connections

Figure 12.3 *Computation in physical neurons and synapses in Boltzmann and back-propagation learning.*

chine, the synapse weights are symmetric although we can show that the algorithm works even when the weights are different in the two directions [Allen and Alspector (1990)]. After settling (the noise mechanisms are not shown in the figure), the weights are adjusted according to the learning rule shown which implies an instantaneous update rather than a batch update.

Panel (d) shows the function of a synapse in back-propagation. The forward path operates as in the Boltzmann case. The backward path does as well although it is attached only to the virtual neurons. The weight is also symmetric in both directions but in the backward path multiplies error rather than activation. The learning rule combines information from both the forward activation and the backward error.

In deciding which algorithm is easier to implement, we chose the Boltzmann machine because there is no easy way in analog VLSI to accurately take the derivative of a neuron's activation function for the backward propagation. Also, back-propagation requires the extra neuron and the virtual paths. Furthermore, the accuracy must be kept high through all layers as the error is propagated backward, making it more suitable for a high precision digital implementation. The Boltzmann learning rule does not require as high precision since it simply balances the clamped and free phases and adjusts weights according to the differences. Note, however, that a contrastive learning rule similar to the Boltzmann machine, with clamped and free phases, can be used to make a back-propagation chip [Morie and Amemiya (1994)]. Since these correlations are measured at equilibrium using the actual physical network, there is no need for an exact model of the activation of each neuron as there is in back-propagation where accurate derivatives are needed to calculate the gradient direction. The difficult part in the Boltzmann machine is making sure the correlation measurements are taken in an unbiased way and at equilibrium. The next section deals with an alternate way of achieving equilibrium.

12.4 Mean-field (deterministic) version of Boltzmann machine

We have referred to the digital states in the Boltzmann model as s_i while the continuous valued activation functions used in back-propagation, we have called y_i. However, Equation 12.10 for the probability of an 'on' state in the Boltzmann machine is similar to Equation 12.15 for the activation in back-prop learning. In fact, using mean-field methods from physics, one can create an approximation to the Boltzmann machine called the mean-field version [Peterson and Anderson (1987)] or deterministic Boltzmann machine [Hinton (1989)]. In this approximation, we replace the discrete neuron states of the Boltzmann machine by its continuous-valued expectation value and use this value as the output activation. Thus,

$$\langle s_i \rangle \equiv \left\langle \frac{1}{1 + e^{-\frac{\sum_j w_{ij} s_j}{T}}} \right\rangle \qquad (12.24)$$

becomes

$$\langle s_i \rangle \approx \frac{1}{1 + e^{-\frac{\sum_j w_{ij} \langle s_j \rangle}{T}}}. \qquad (12.25)$$

Identifying $\langle s_i \rangle$ with the activation y_i, we have

$$y_i \approx \frac{1}{1 + e^{-\frac{\sum_j w_{ij} y_j}{T}}} = \frac{1}{1 + e^{-\beta x_i}} \qquad (12.26)$$

which is the same activation function as that typically used in back-propagation except for the 'gain' term $\beta = \frac{1}{T}$.

The mean field approximation also assumes that the expectation of correlations can be factored, which is often a poor approximation:

$$\langle s_i s_j \rangle \approx \langle s_i \rangle \langle s_j \rangle = y_i y_j. \qquad (12.27)$$

Thus the Boltzmann machine learning rule

$$\Delta w_{ij} = -\eta \left(\langle s_i s_j \rangle^+ - \langle s_i s_j \rangle^- \right) \qquad (12.28)$$

becomes

$$\Delta w_{ij} = -\eta \left(y_i^+ y_j^+ - y_i^- y_j^- \right). \qquad (12.29)$$

This model has the advantage that the long annealing time to arrive at equilibrium in the Boltzmann machine is greatly reduced. Rather than the temperature T being gradually lowered (noise reduced), the gain, β , is increased from near zero to a high value. This is computationally more efficient on a computer (fewer cycles to reach equilibrium) and is also very natural for VLSI, as we shall see.

12.5 Electronic implementation of a Boltzmann machine

Our model incorporates both the stochastic and deterministic versions of the Boltzmann machine. The electronic neuron has activation

$$y_i = f[\beta(x_i + \nu_i)] \qquad (12.30)$$

where f is a monotonic non-linear function similar to the sigmoid and ν_i is an additive noise, independent for each neuron. The noise has a distribution close to a zero-mean gaussian taken from our electronic implementation [Alspector, Gannett, Haber, Parker and Chu (1991)]. This closely approximates the Boltzmann distribution of Equation. 12.10. Actually, the integral of a gaussian is the error function $\frac{1}{2}(1 + erf(\beta x_i))$, which, properly scaled, is within a few percent of a sigmoid over its range. The noise can be lowered as annealing proceeds. If the gain, β , is high, the neuron acts like the step function of the original Boltzmann model.

To perform mean field annealing, we keep the noise at zero and change the gain, β , of the neuron amplifier. We can anneal using both methods simultaneously if we wish. The time average of the neuron activation with noise jitter is precisely the activation of the mean-field approach. As the noise jitter is reduced, the time average sharpens. In the mean-field, this is equivalent to the neuron having higher gain. Thus, reducing the noise in the stochastic Boltzmann machine is like increasing the neuron gain in the mean-field version.

The network is annealed (by either or both methods) in the clamped (+) and free (-) phases and weights are adjusted using

$$\Delta w_{ij} = \frac{1}{2}\left(sgn(y_i y_j)^+ - sgn(y_i y_j)^-\right) \qquad (12.31)$$

by measuring the instantaneous correlations after annealing. The weights are stored digitally at each synapse and range from -15 to +15 in value. We can decay selected weights by reducing their magnitude by one.

12.5.1 The learning microchips

We have designed and fabricated two kinds of experimental prototype learning microchips. One contains neurons and synapses and another contains only synapses meant for interconnecting the

Figure 12.4 *Microphotograph of the chip containing 32 neurons and 992 connections.*

neuron-containing chips. One contains 32 neurons and 992 connections (496 bidirectional synapses) (see microphotograph of the chip in Figure 12.4). There is a noise generator which supplies 32 uncorrelated pseudo-random noise sources [Alspector, Gannett, Haber, Parker and Chu (1991)] to the neurons to their left. These noise sources are summed in the form of current along with the weighted post-synaptic signals from other neurons at the input to each neuron in order to implement the simulated annealing process of the stochastic Boltzmann machine. The neuron amplifiers implement a non-linear activation function which has variable gain to provide for the gain sharpening function of the mean-field technique. The range of neuron gain can also be adjusted to allow for scaling in summing currents due to adjustable network size.

Most of the area is occupied by the synapse array. Each synapse digitally stores a weight ranging from -15 to +15 as 4 bits plus a sign. It multiples the voltage input from the presynaptic neuron by this weight to output a current. One conductance direction can be disconnected so that we can experiment with asymmetric networks [Allen and Alspector (1990)]. Although the synapses can have their weights set externally, they are designed to be adaptive.

They store correlations, in parallel, using the local learning rule of Equation. 12.31 and adjust their weights accordingly. The synapse counts as correlations when the two neurons it connects are both either on (above 2.5 volts) or off (below 2.5 volts) and therefore a neuron state range of -1 to 1 is assumed by the digital learning processor in each synapse on the chip.

In order to interconnect the neuron-containing chips to form larger artificial neural systems, we need a chip which contains only synapses. The nodes which sum current from synapses for net input into a neuron in the neuron-containing chips are available externally for connection to other chips and for external clamping of neurons or other external input. We designed and fabricated such a chip for our experimental prototype microsystem. It contains an array of 32 by 32 (1024) learning synapses for interconnecting the neuron containing chips (see photo in Figure 12.5). We have integrated these chips in a 3 chip experimental prototype system containing 2 neuron chips and 1 synapse chip. This system is equivalent to a fully interconnected 64 neuron learning system and has 4 times the number of synapses that our single 32 neuron learning chip has.

12.5.2 Learning experiments

We have done several experiments on a system consisting of only a single learning chip [Alspector, Jayakumar and Luna (1992)]. To study learning as a function of problem size, we chose the parity and replication (identity) problems. This facilitates comparisons with our previous simulations [Alspector, Allen, Hu and Satyanarayanna (1988)]. The parity problem is the generalization of exclusive-OR for arbitrary input size. It is difficult because the classification regions are disjoint with every change of input bit, but it has only one output. The goal of the replication problem is for the output to duplicate the bit pattern found on the input after being encoded by the hidden layer. Note that the output bits can be shifted or scrambled in any order without affecting the difficulty of the problem. There are as many output neurons as input. For the replication problem, we chose the hidden layer to have the same number of neurons as the input layer, while for parity we chose the hidden layer to have twice the number as the input layer. Interested readers are referred to [Alspector, Jayakumar and Luna (1992)].

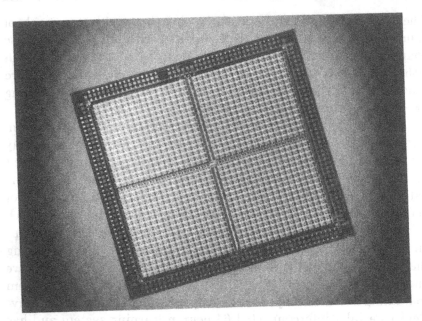

Figure 12.5 *Microphotograph of the* 32×32 *synapse array chip used for interconnecting the neuron chips of Figure 12.4*

We have done on-chip learning experiments using noise and gain annealing for parity and replication up to 8 input bits, nearly utilizing all the neurons on a single chip. To judge scaling behavior in these early experiments, we note the number of patterns required until no further improvement in percent correct is visible by eye. This scales roughly as an exponential in number of inputs for learning on chip just as it did in simulation [Alspector, Allen, Hu and Satyanarayanna (1988)] since the training set size is exponential. Overall, the training time per pattern on-chip is quite similar to our simulations. However, in real-time, it can be about 100,000 times as fast for a single chip and will be even faster for multiple chip systems. The speed for either learning or evaluation is roughly 100 million connections per second per chip.

12.6 Building a system using the learning chips

12.6.1 VME system and application areas

In order to rapidly enable applications prototyping, we have integrated our learning chips into a standard computer system. It functions as a co-processor in a Sun Sparcstation(TM) through the VME bus interface and is controlled by a user-friendly, graphical interface. The system uses the previously described cascadable analog learning chips [Alspector, Jayakumar and Luna (1992)]. A three-chip network consisting of 64 neurons and about 1000 adaptive synapses has been successfully tested. The system is functioning as a general purpose test bed for running neural network applications.

12.6.2 The co-processor system

The system consists of the following components: 1) Controller board, 2) Analog and digital pattern generation/acquisition boards and 3) Neural network boards. These boards constitute a VME slave sub-system that can be easily interfaced with any high - performance computing platform.

1. The Controller board: This board handles the VME interface between the neural network and the host. Address translation circuitry translates higher level commands into control signals for the neural net chips. The control functions performed by this board are weight read/writes, neural network topology setup, noise and gain annealing schedule generation for relaxation of the network, learning control and seeding the noise generator. This controller can support up to 16 neural network chips.

2. The pattern generation/acquisition system: This system consists of a digital board for fast 64 bit-digital input pattern presentation and 32 bit readouts from the neural net chips. Off-the-shelf, fast multi-channel A/D and D/A conversion boards enable 32 channel analog pattern presentation/acquisition to and from the boards containing the neural network chips. The analog I/O signals are routed to the neural network boards through a private analog bus which serves as the main 'pattern' bus for the network.

3. The neural network board: This board houses the neural network chips (see photo in Figure 12.6). Presently, we have a three-chip board but any number of NN boards can be used limited only by connectivity constraints. This board is shielded from the VME bus to reduce the impact of high speed switching noise and is connected to the host through the controller and pattern cards via the control and pattern buses.

Figure 12.6 *Photograph of the VME Boltzmann neural network board.*

12.6.3 System software

The system is controlled by our in-house neural network simulation
tool, 'Braincore'. This is a versatile, X-windows based, graphics
oriented, C language software package designed to simulate various
neural network algorithms. The basic graphics interface controls
the operation of the hardware through a series of Unix system calls
to the co-processor. The user can set up a network topology, control
parameters such as annealing schedules and neuron gains and point
to a pattern file which contains the training and test vectors. The
software translates the virtual network map to physical connections
on the cascaded chip network, downloads the annealing schedules
to the on-board RAM-DACs, puts the patterns on the pattern
bus and activates the learning cycle. To the user, the hardware is
totally transparent and just looks like another simulation module
to be invoked, except now it runs on the fast co-processor hardware.

12.6.4 Experimental results

The system has been designed for general purpose use so as to aid
the speedy deployment of neural network techniques in real-time

hardware after extensive study and simulation of the problem by conventional neural network methods but before any custom hardware has to be built. Currently we are interested in the problem of fast adaptive equalization of highly non-linear channels, such as wireless channels encountered in Personal Communications Systems. The co-processor hardware can validate the implementation schemes in a fairly realistic environment.

We have run the equalization task for a noisy channel but with the learning done using our hardware system. The software only presents bit patterns to the hardware and displays the results. The network corrects nearly all errors after a few thousand bits for non-linear channels and a few tens of bits for linear channels. Interested readers are referred to [Jayakumar and Alspector (1994)].

12.7 Other applications

We expect to be able to prototype many other applications on this system because of its many modes of operation. These include 1) classification (either in feedback or in feed-forward), 2) optimization using the settling mechanisms of noise and gain control (this includes content-addressable memory as a special case), and 3) learning for real-time adaptation.

For classification tasks, the weights can be learned by the actual system beforehand and used in the feedback mode (which is how the weights were learned). This might be used for word-spotting where there is user-dependent speech recognition training before use and learning time is a bottleneck. The system would classify by settling using gain or noise variation which requires 10-100 microseconds per pattern. This should be sufficient for most classification applications. It may, however, be desirable to use a feed-forward system for classification in those cases where the weights have been learned by an off-line technique such as back-propagation. In that case, the system has the capability of downloading fixed weights from a file and removing one direction of its bi-directional synapse from operation. The system will not need to settle its feedback paths and will be faster (about 3 microseconds per pattern). This mode might be used for user verification by speech, face [Solheim, Payne and Castain (1992)], or handwriting where weights will be downloaded for each user. Another application might be fault identification and network management [Goodman, Miller and Latin (1988)] where sensor information from other parts of the network is coming so quickly that the extra speed is needed and off-line learning is sufficient. A particularly demanding application is pattern classification in high-energy physics experiments where the data rate is extremely high.

For optimization tasks, one can use the built-in settling features of the feedback network with downloaded weights appropriate for the task. This could be used for switching and routing problems using the techniques of T. X. Brown (1989) by prototyping winner-take-all circuits with inhibitory feedback connections. One could also use the Hopfield style optimization technique for crossbar switching as advocated by Marrakchi and Troudet (1989). Other problems where these techniques can be applied include dynamic bandwidth allocation in wireless networks [Kunz (1991)], and packet routing [Ali and Nguyen (1989)]. Of course, our previously described CAM settles to learned codewords. This can be used for vector quantization and other coding techniques. Another example of an optimization task is CDMA multiuser detection.

For real-time adaptation, this system uses its capabilities fully. We have already described an application in channel equalization which is best suited for noisy or non-linear channels [Chen, Gibson and Cowan (1990)]. This technique is also applicable to noise or echo-cancellation filtering.

Another application is access control in packet networks [Hiramatsu (1990)]. For this, the system would learn the highly nonlinear nature of the packet network given the quality of service requirements of a call-request. This will have to be learned on-line for different parts of the network, different user profiles, etc. It is also likely to change as new equipment or new services are added to the network requiring a continuously adaptive system.

Personal user identification is different from verification in that new or unknown users may request service and a profile must be generated on the spot. This requires high-speed learning of voice, face [Solheim, Payne and Castain (1992)], handwriting, etc. It may also be necessary for some applications, like distance-learning, to identify a face in a crowd or in new situations. A learning system has advantages here.

For video data compression [Cottrell, Munro and Zipser (1987)] using vector quantization or predictive coding techniques, a learning system can learn the characteristics of the source data and adapt the code-book for changing scenes. For multi-user communication in a spread-spectrum CDMA wireless communication sys-

tem, adaptive techniques can be used to solve the near-far problem
[Aazhang, Paris and Orsak (1992)].

References

Aazhang, B., Paris, B.P. and Orsak, G.C. (1992) Neural networks for multiuser detection in CDMA communications. *IEEE Trans. on Communications*, **40**, 1212–1222.

Abu-Mostafa, Y.S. (1993) Hints and the VC Dimension. *Neural Computation*, **5**, 278–288.

Ackley, D.H., Hinton, G.E. and Sejnowski, T.J. (1985) A learning algorithm for Boltzmann machines. *Cognitive Science*, **9**, 147–169.

Ali, M.M. and Nguyen, H.T. (1989) A Neural Network Implementation of an Input Access Scheme in a High Speed Packet Switch", *Proceedings of the 1989 Global Telecommunications Conference*, 1192-1196.

Allen, R.B. and Alspector, J. (1990) Learning of Stable States in Stochastic Asymmetric Networks. *IEEE Trans. Neural Networks*, **1**, 233–238.

Alspector, J., Allen, R.B., Hu, V. and Satyanarayanna, S. (1988) Stochastic learning networks and their electronic implementation. *Proceedings of the conference on Neural Information Processing Systems, Denver, CO, Anderson, D. Ed. New York, NY: Am. Inst. of Phys.*, 9–21.

Alspector, J., Gupta, B. and Allen, R.B. (1989) Performance of a stochastic learning microchip. *Advances in Neural Information Processing Systems D.S. Touretzky, Editor, Morgan Kaufmann Publishers*, **1**, 748–760.

Alspector, J., Allen, R.B., Jayakumar, A., Zeppenfeld, T. and Meir, R. (1991) Relaxation Networks for Large Supervised Learning Problems. *Advances in Neural Information Processing Systems, R. P. Lippmann, J. E. Moody and D. S. Touretzky Editors, Morgan Kaufmann Publishers*, **3**, 1015–1021.

Alspector, J., Gannett, J.W., Haber, S., Parker, M.B. and Chu, R. (1991) A VLSI-Efficient Technique for Generating Multiple Uncorrelated Noise Sources and Its Application to Stochastic Neural Networks. *IEEE Trans. Circuits and Systems*, **38**, 109.

Alspector, J., Jayakumar, A. and Luna, S. (1992) Experimental Evaluation of Learning in a Neural Microsystem. *Advances in Neural Information Processing Systems, ed. Moody, J.E., Hanson, S.J., Lippmann, R.P ., Morgan Kauffman Publishers*, **4**, 871–878.

Alspector, J., Meir, R., Yuhas, B. and Jayakumar, A. (1993) A Parallel Gradient Descent Method for Learning in Analog VLSI Neural Networks. *Advances in Neural Information Processing Systems, S.J. Hanson, J.D. Cowan and C.L. Giles Editors, Morgan Kaufmann Publishers*, **5**, 836–844.

Anderson, J.A., Silverstein, J.W., Ritz, S.A. and Jones, R.S. (1977) Distinctive features, categori cal perception, and probability learning: Some applications of a neural model. *Psych. Rev.*, **84**, 413–451.

Bastiaansen, C.A.A., Wouter J.G.D., Schouwenaars, H.J. and H.A.H. Termeer (1991). A 10-b 40mhz $0.8\mu m$ CMOS current–output D/A converter. *IEEE Journal of Solid-State Circuits*, **26(7)**, 917–921.

Berglund, C.N. (1971) Analog performance limitations of charge–transfer dynamic shift registers. *IEEE Journal of Solid-State Circuits*, **6**, 391–394.

Berglund, C.N. and Thornber, K.K. (1973) Incomplete transfer in charge–transfer devices. *IEEE Journal of Solid-State Circuits*, **8**, 108–116.

Binder, K. (1978) (ed.) The Monte-Carlo Method in Statistical Physics. *Springer-Verlag, NY.*

Brown, T.X. (1989) Neural Networks for Switching. *IEEE Communications Magazine*, **27:11**, 72–81.

Burr, J.B. and Shott, J. (1994) A $200mW$ Self-Testing Encoder/Decoder using Stanford Ultra-Low-Power CMOS. *IEEE International Solid State Circuits Conference, Digest of Technical Papers*, 84–85.

Castello, R., Caviglia, D.D., Franciotta, M. and Montechhi, F. (1991) Self-Refreshing Analogue Memory Cell for Variable Synaptic Weights. *Electronics Letters*, **27**, 1871–1873.

Castro, H.A. and Sweet, M.R. (1993) Radiation Exposure Effects on the Performance of an Electrically Trainable Artificial Neural Network. *IEEE Transactions on Nuclear Science*, **40(6)**, 1575–1583.

Castro, H.A., Tam, S.M. and Holler, M.A. (1993) Implementation and Performance of an analogue Nonvolatile Neural Network. *Analogue Integrated Circuits and Signal Processing*, **4(2)**, 97–113.

Cauwenberghs, G. (1993) A Fast Stochastic Error-Descent Algorithm for Supervised Learning and Optimisation. *Advances in Neural Information Processing Systems, S.J. Hanson, J.D. Cowan and C.L. Giles Editors, Morgan Kaufmann Publishers*, **5**, 244–251.

Cauwenberghs, G. (1994a) A Learning Analog Neural Network Chip with Continuous-Time Recurrent Dynamics. *Advances in Neural Information Processing Systems, J.D. Cowan, G. Tesauro, and J. Alspector Editors, Morgan Kaufmann Publishers*, **6**, 858–865.

Cauwenberghs, G. (1994) Analog VLSI Autonomous Systems for Learning and Optimization. *PhD thesis, California Institute of Technology, Pasadena, California*

Chen, S., Gibson, G.J. and Cowan, C.F.N. (1990) Adaptive channel equalisation using a polynomial-perceptron structure. *IEE Proceedings*, **137**, 257–264.

Coggins, R.J., Jabri, M.A. and Pickard, S.J. (1993) A Comparison of Three On Chip Neuron Designs For a Low Power Analogue VLSI MLP. *Microneuro '93, Edinburgh, Scotland, UK.*, 97–103.

Coggins, R.J. and Jabri, M.A. (1994) WATTLE: A Trainable Gain Analogue VLSI Neural Network, *Advances in Neural Information Processing Systems, J.D. Cowan, G. Tesauro, and J. Alspector Editors, Morgan Kaufmann Publishers*, **6**, 874–881.

Coggins, R.J., Jabri, M.A., Flower B.F. and Pickard, S.J. (1995) A Hybrid Analog and Digital Neural Network for Intracardiac Morphology Classification. *IEEE Journal of Solid State Circuits*, **30:5**, 542–550.

Coggins, R.J., Jabri, M.A., Flower B.F. and Pickard, S.J. (1995a) ICEG Morphology Classification using an Analogue VLSI Neural Network. *Advances in Neural Information Processing Systems, G. Tesauro, D.S. Touretzky and T.K. Leen Editors, Morgan Kaufmann Publishers*, **7**.

Coggins, R.J., Jabri, M.A. Flower B.F. and Pickard, S.J. (1995b) A Low Power Network for On-Line Diagnosis of Heart Patients. *IEEE Micro Magazine*, **15:3**, 18–25.

Cohen, M.A. and Grossberg, S. (1983) Absolute stability of global pattern formation and parallel memory storage by competitive neural networks. *Trans. IEEE SMC*, **13**, 815.

Cottrell, G.W., Munro, P. and Zipser, D. (1987) Image Compression by Back Propagation: An Example of Extensional Programming. *ICS Report 8702, University of California, San Diego*.

Dembo, A. and Kailath, T. (1990) Model-Free Distributed Learning, *IEEE Transactions on Neural Networks*, **1:1**, 58-70.

Enz, C.C., Krummenacher, F. and Vittoz, E.A. (1994) An Analytical MOS Transistor Model Valid in All Regions of Operation and Dedicated to Low Voltage and Low Current Applications. *Analog Integrated Circuits and Signal Processing, November 1994.*

Flower, B.F. and Jabri, M.A. (1993a) Summed Weight Neuron Perturbation: an O(N) improvement over Weight Perturbation. *Morgan Kauffmann Publishers, NIPS5*, **5**, 212–219.

Flower, B.F. and Jabri, M.A. (1993b) The Implementation of Single and Dual Transistor VLSI Analogue Synapses. *Proceedings of the Third International Conference on Microelectronics for Neural Networks*, 1–10.

Flower, B.F. and Jabri, M.A. (1994) A Learning Analogue VLSI Neural Network. *SEDAL, Department of Electrical Engineering, University of Sydney.*

Flower, B.F. and Jabri, M.A. and Pickard, S. (1994) An Analogue On-

Chip Supervised Learning Implementation of an Artificial Neural Network. *submitted to IEEE Transactions on Neural Networks.*

Franca, J.E. and Tsividis, Y.P. (1994) Design of Analog and Digital VLSI Circuits for Telecommunications and Signal Processing. *Prentice-Hall.*

Geiger, R.L., Allen, P.E. and Strader, N.R. (1990) VLSI Design Techniques for Analog and Digital Circuits. *McGraw-Hill*

Goodman, R.M., Miller, J. and Latin, H. (1988) NETREX: A real time network management expert system. *IEEE Globecom Workshop on the Application of Emerging Technologies in Network Operation and Management.*

Graf, H.P. et al., (1986) VLSI Implementation of a Neural Network Memory with Several Hundreds of Neurons. *Proceedings of the Conference on Neural Networks for Computing, Amer. Inst. of Phys.* 182.

Graf, H.P. and Henderson, D. (1990) A reconfigurable CMOS neural network. *IEEE International Solid-State Circuits Conference, Digest of Technical Papers.*

Graf, H.P. and Cosatto, E. (1994) Address block location with a neural net system. *Neural Information Processing Systems, J. Alspector, J. Cowan and G. Tesauro editor,* **6**, 785–792.

Gray, P.R. and Meyer, R.G. (1993) Analysis and Design of Analog Integrated Circuits. *3rd Ed, Wiley.*

Gray, P.R. and Napp, R.R. (1994) A-D Conversion Techniques for Telecommunications Applications. *Design of Analog and Digital VLSI Circuits for Telecommunications and Signal Processing. J E Franca and Y P Tsividis (Eds), Prentice-Hall*

Gregorian, R., Martin, K.W. and Temes, G.C. (1983) Switched Capacitor Circuit Design. *IEEE Journal of Solid State Circuits,* **71(8)**, 941–966.

Gregorian, R. and Temes, G.C. (1986) Analog MOS Integrated Circuits for Signal Processing, *John Wiley and Sons.*

Grossberg, S. (1969) On learning and energy-entropy dependence in recurrent and nonrecurrent signed networks. *Jour. Stat. Phys.,* 1, 319.

Grossberg, S. (1973) Contour enhancement, short term memory, and constancies in reverberating neural networks. *in Studies in Applied Mathematics, LII,* **213**, *MIT Press.*

Hammerstrom, D. (1993) Neural networks at work. *IEEE Spectrum,* **June**, 26–32.

Haskard, M.R. and May, I.C. (1988) Analog VLSI Design: nMOS and CMOS. *Prentice-Hall*

Haykin, S. (1994) Neural Networks, A comprehensive foundation. *Macmillan College Pulishing Company.*

Hebb, D.O. (1949) The Organization of Behavior. *Wiley, NY.*

Henderson, P.C.B. (1990) Analogue Correlators in Standard MOS Technology. *PhD Thesis, University of Sydney.*

Hertz, J., Krogh, A. and Palmer, R.G. (1991) Introduction to the Theory of Neural Computation. *Addison Wesley.*

Hinton, G. E. (1989) Deterministic Boltzmann Learning Performs Steepest Descent in Weight Space. *Neural Computation*, 1, 143–150.

Hiramatsu, A. (1990) ATM Communications Network Control by Neural Network. *IEEE Transactions On Neural Networks*, 1:1.

Hochet, B. (1989) Multivalued MOS Memory for Variable Synapse Neural Networks. *Electronics Letters*, 25:10.

Hochet, B., Peiris, V., Abdo, S. and Declercq, M. (1991) Implementation of a Learning Kohonen Neuron Based on a New Multilevel Storage Technique. *IEEE Journal of Solid-State Circuits*, 26:3.

Hoehfeld, M. and Fahlman, S.E (1991) Learning with limited precision using the cascade–correlation algorithm, *Technical report CMU–CS–91–130*, School of computer science, Pittsburgh, USA.

Hoff, M.E. and Townsend, M. (1979) An analog input-output Microprocessor. *IEEE International Solid-State Circuits Conference, Digest of Technical Papers.*

Hollis, P.W., (1990) The effects of precision constraints in a backpropagation learning network. *Neural Computation*, MIT Press, 2:3, 363–373.

Hopfield, J.J. (1982) Neural networks and physical systems with emergent collective computational abilities. *Proc. Natl. Acad. Sci. USA*, 79, 2554–2558

Jabri, M., Pickard, S., Leong, P., Rigby, G., Jiang, J., Flower, B., and Henderson, P. (1991) VLSI Implementation of Neural Networks with Application to Signal Processing. *Proceedings of IEEE International Symposium of Circuits and Systems, Singapore*, 1275–1278.

Jabri, M.A. and Flower, B.F. (1992) Weight perturbation: An optimal architecture and learning technique for analog VLSI feedforward and recurrent multilayer networks. *IEEE Transactions on Neural Networks*, 3:1, 154–157.

Jabri, M.A. and Flower, B. (1993) Practical Performance of Analogue Multi-layer Perceptron Training Algorithms. submitted to *IEEE Transactions on Neural Networks.*

Jabri, M.A., Pickard, S., Leong, P.H.W. and Xie, Y. (1993) Algorithms and Implementation Issues in Analog Low Power Learning Neural Network Chips. *International Journal on VLSI Signal Processing, U.S.A.* , 6:2, 67–76.

Jabri, M.A., Tinker, E.A. and Leerink, L. (1994) MUME – A Multi-Net-Multi-Architecture Neural Simulation Environment. *Neural Network Simulation Environments, Kluwer Academic Publications.*

Jayakumar, A. and Alspector, J. (1994) A Neural Network based Adaptive Equalizer for Digital Mobile Radio. *in Government Microelec-*

tronics Applications Conference, (San Diego, CA,).

Kirkpatrick, S., Gelatt, C.D., and Vecchi, M.P. (1983) Optimization by simulated annealing. *Science*, **220**, 671–680.

Kirkpatrick, S. and Sherrington, D. (1978) Infinite-ranged models of spin-glasses. *Phys. Rev.* , **17**, 4384–4403.

Kohonen, T. (1977) Associative memory - A system-theoretic approach. *Springer-Verlag, Berlin.*

Kunz, D. (1991) Channel Assignment for Cellular Radio Using Neural Networks. *IEEE Trans. on Vehicular Technology*, **40:1**, 188–193.

Kusumoto, K. and et. al. (1993) A 10bit 20Mhz 30mW Pipelined Interpolating ADC. *IEEE Solid State Circuits Conference, Digest of Technical Papers*, 62–63.

Leong, P.H.W. (1992a) Arrhythmia Classification Using Low Power VLSI. *PhD Thesis, University of Sydney.*

Leong, P.H.W. and Jabri, M. (1992b) A VLSI neural network for arrhythmia classification Classification System. *Proceedings of the International Joint Conference on Neural Networks, Baltimore, Maryland USA*, **2**, 678–683.

Leong, P.H.W. and Jabri, M. (1992c) MATIC - An Intracardiac Tachycardia Classification System. *Pacing & Clinical Electrophysiology*, **15**, 1317–1331.

Leong, P.H.W. and Vasimalla, J.G. (1992) JIGGLE User's Manual. *Sedal Technical Report PL-0892*, Sydney University Electrical Engineering.

Leong, P.H.W. and Jabri, M.A. (1993a) A Low Power Analogue Neural Network Classifier Chip. *San Diego, USA, Proceedings of the IEEE Custom Integrated Circuits Conference*, 4.5.1–4.5.4.

Leong, P.H.W., and Jabri, M.A. (1993b) Kakadu - A Low Power Analogue Neural Network. *International Journal of Neural Systems*, **4:4**, 381–394.

Lin, D., Dicarlo, L.A. and Jenkins, J.M. (1988) Identification of Ventricular Tachycardia using Intracavitary Electrograms: analysis of time and frequency domain patterns. *Pacing and Clinical Electrophysiology*, **11**, Part I, 1592–1606.

Lyle, J.D (1992) Sbus, Information, Application and Experience. *Springer Verlag.*

Marrakchi, A.M. and Troudet, T. (1989) A Neural Net Arbitrator for Large Crossbar Packet-Switches. *IEEE Transactions on Circuits and Systems*,**36:7**, 1039–1041.

McCulloch, W.S. and Pitts, W.H. (1943) A logical calculus of ideas immanent in nervous activity. *Bulletin of Mathematical Biophysics*,5(1), 115.

Mead, C. (1989) Analog VLSI and Neural Systems. *Addison-Wesley.*

Metropolis, N., Rosenbluth, A., Rosenbluth, M., Teller, A., and Teller, E. (1953) Equation-of-state calculations by fast computing machines.

Jour. Chem. Phys., **6**, 1087.

Minsky, M. and Papert, S. (1969) Perceptrons. *MIT Press, Cambridge, MA.*

Morie, T. and Amemiya, Y. (1994) An All-Analog Expandable Neural Network LSI with On-Chip Backpro pagation Learning. *IEEE Jour. Solid-State Circuits*, **29**, 1086.

Motchenbacher, C.D. and Fitchen, F.C. (1973) Low Noise Electronic Design. *Wiley, NY,* 9.

Moopenn, A., Duong, T. and Thakoor, A.P. (1990) Digital-Analog Hybrid Synapse Chips for Electronic Neural Networks. *in Advances in Neural Information Processing Systems, D. Touretzky Editor, Morgan Kaufmann Publishers,* 769–776.

Mueller, P., Van der Spiegel, J., Blackman, D., Chiu, T., Clare, T., Donham, C., Hsieh, T.P. and Loinaz, M. (1989) Design and Fabrication of VLSI Components for a General Purpose Analog Neural Computer. *Analog VLSI Implementation of Neural Systems, C. Mead and M. Ismail, Kluwer,* 135–169.

Murray, A. F. (1991) Analogue Noise-Enhanced Learning in Neural Network Circuits. *Electronic Letters*, 27, 1546–1548.

Murray, A. and Tarassenko, L. (1994) Analogue Neural VLSI — A pulse stream approach. *Chapman & Hall*

Pedroni, V.A., (1994) Self-Contained Error-Compensated N-valued Memory for Neural Applications *Fourth International Conference on Microelectronics for Neural Networks and Fuzzy Systems, Turin, Italy, September,* 152–155

Peiris, V. (1994) Mixed Analog Digital VLSI Implementation of a Kohonen Neural Network. *PhD thesis No. 1295, Swiss Federal Institute of Technology (EPFL) Lausanne*

Peterson, C. and Anderson, J.R. (1987) A Mean Field Learning Algorithm for Neural Networks. *Complex Systems*, **1:5**, 995–1019.

Pickard, S.J., Jabri, M.A., Leong, P.H.W, Flower, B.G. and Henderson, P.C.B. (1992) Low Power Analogue VLSI Implementation of a Feed Forward Neural Network. *Proceedings of the Third Australian Conference on Neural Networks,* 88–91.

Pickard, S.J. (1993a) Common Mode Compensating Neuron. *SEDAL, Dept. ELec. Eng., University of Sydney, February, SEDAL Internal Report.*

Pickard, S.J. (1993b) Further Work on CoMCoN. *SEDAL, Dept. ELec. Eng., University of Sydney, March, SEDAL Internal Report.*

Pickard, S.J., Jabri, M.A. and Flower, B.G. (1993) VLSI Neural Network for Classification in Very Low Power Applications. *SEDAL ICEG-RPT-008-93, Department of Electrical Engineering, University of Sydney.*

Raffel, J., Mann, J., Berger, R., Soares, A. and Gilbert, S. (1987) A

Generic Architecture for Wafer-Scale Neuromorphic Systems. *in Proceedings of IEEE International Conference on Neural Networks, San Diego, USA, June,* 135–169

Rosenblatt, F. (1961) Principles of Neurodynamics: Perceptrons and the theory of brain mechanism s. *Spartan Books, Washington, D.C..*

Rumelhart, D.E., Hinton, G.E. and Williams, R.J. (1986) Learning internal representations by error propagation. *in Parallel Distributed Processing: Explorations in the Microstructure of Cognitioni, vol.1: Foundations, Rumelhart, D.E. and McClelland, J.L. (eds.), MIT Press, Cambridge, MA.* 318.

Rumelhart, D.E., McClelland, J.L. and the PDP Research Group, (1986) Parallel Distributed Processing. Vol. 1, *MIT Press.*

Sackinger, E., Boser, B.E., Bromley, J., LeCun, Y. and Jackel, L.D. (1992) Application of the ANNA Neural Network Chip to High-Speed Character Recognition. *IEEE Trans. on Neural Networks,* **3**, 498–505.

Sejnowski, T.J. (1981) Skeleton filters in the brain. *in Parallel Models of Associative Memory, G. Hinton & J.A. Anderson (eds.), Erlbaum, Hillsdale, NJ.,* 189–212.

Sivilotti, M., Emerling, M. and Mead, C. (1985) A Novel Associative Memory Implemented Using Collective Computation. *Proceedings of the 1985 Chapel Hill Conference on Very Large Scale Integration,* 329.

Solheim, I., Payne, T.L. and Castain, R.C. (1992) The Potential in Using Backpropagation Neural Networks for Facial Verification Systems. *WINN-AIND, Auburn, AL.*

Suarez, R.E., Gray, P.R. and Hodges, D.A. (1975) All MOS Charge Redistribution A to D Conversion Techniques. *IEEE Journal of Solid State Circuits.,* **SC-10(6)**, 379–385.

SUN Microsystems inc. (1990) SBUS Specification B.0.

Sze, S. (1984) Physics and Technology of Semiconductor Devices. *Wiley*

Tam, S.M., Gupta, B., Castro, H.A. and Holler, M.A. (1990) Learning on an Analog VLSI Neural Network Chip. *IEEE International Conference of Systems, Man and Cybernetics,* 701–703.

Tarassenko, L., Tombs, J. and Cairns, G. (1993) On-chip learning with analogue VLSI neural networks, *International journal of neural systems,* **4.4**, 419-426.

Throne, R.D., Jenkins, J.M. and DiCarlo, L.A. (1991) A Comparison of Four New Time-Domain Techniques for Discriminating Monomorphic Ventricular Tachycardia from Sinus Rhythm Using Ventricular Waveform Morphology. *IEEE Transactions on Biomedical Engineering,* **38**, 561–571.

Tinker, E.A. (1992) The SPASM Algorithm for Ventricular Lead Timing and Morphology Classification. *SEDAL ICEG-RPT-016-92, Department of Electrical Engineering, University of Sydney.*

Tsividis, Y.P. and Antognetti, A. (1985) Design of MOS VLSI Circuits for Telecommunications. *Prentice-Hall.*

Tsividis, Y.P. (1988) Operation and Modeling of the MOS Transistor. *McGraw-Hill.*

Vittoz, E.A., Oguey, H., Maher, M.A., Nys, O., Dijkstra, E. and Chevroulet, M. (1991) Analog Storage of Adjustable Synaptic Weights. *in VLSI Design of Neural Networks, Norwell MA: Kluwer Academic,* 47-63.

Vittoz, E.A. (1991) Intensive Summer Course on CMOS and BiCMOS VLSI Design'91. *EPFL Lausanne Switzerland.*

Vittoz, E.A. (1994) Micropower Techniques. *Design of Analog and Digital VLSI Circuits for Telecommunications and Signal Processing. J E Franca and Y P Tsividis (Eds), Prentice-Hall*

Weste, N. and Eschragian, K. (1993) Principles of CMOS VLSI Design. *(2nd Ed) Addison-Wesley.*

Widrow, B. and Lehr, M.A. (1990) 30 years of adaptive neural networks: Perceptron, madaline, and backpropagation. *Proceedings of the IEEE,* **78(9)**, 1415-1442.

Widrow, G. and Hoff, M.E. (1960) Adaptive switching circuits. *Inst. of Radio Engineers, Western Electric Show and Convention, Convention Record,* **Part 4**, 96–104.

Xie, Y. and Jabri, M.A. (1992a) On the Training of Limited Precision Multi-layer Perceptron. *Proceedings of the International Joint Conference on Neural Networks,* **III**, 942–947.

Xie, Y. and Jabri, M.A. (1992b) Analysis of the effects of quantization in multilayer neural net works using a statistical model. *IEEE Transactions on Neural Networks,* **3(2)**, 334–338.

Index